T0180434

Computational Artifacts

Raymond Turner

Computational Artifacts

Towards a Philosophy of Computer Science

 Springer

Raymond Turner
School of Computer Science
& Electronic Engineering
University of Essex
Colchester, UK

ISBN 978-3-662-58559-7   ISBN 978-3-662-55565-1   (eBook)
https://doi.org/10.1007/978-3-662-55565-1

Printed on acid-free paper

This Springer imprint is published by the registered company Springer-Verlag GmbH, DE part of Springer Nature.
The registered company address is: Heidelberger Platz 3, 14197 Berlin, Germany

Preface

This book has been some time coming. There are several reasons for this.

For some of the subject matter I had a clear conception of the central questions and concerns, but there were other aspects for which my ideas were barely worked out. Preliminary work on these other topics took considerable time. Not only had I not given them much thought, but neither had anyone else. More generally, the literature on the philosophy of computer science, at least in my conception of it, was sparse.

Secondly, I had no overriding conceptual framework. This made it difficult even to organize the material that I did have. Eventually this was provided by the notion of a *technical artifact*, and with this the speed of writing increased a little. But there was still a major concern: how to fit all the topics which I now felt should be part of any such book within this framework. Here the literature on the philosophy of technology provided some real help, but there was still considerable work to do on reinterpreting matters in computational terms.

Thirdly, the book must appeal, and be accessible to, philosophers as well as computer scientists. The subject matter of computer science is highly technical in nature, and part of any such book, at least at this stage in the development of the subject, must provide an exposition of the relevant core material. Books on computer science are written for a technically minded audience who have the right mathematical and engineering backgrounds. In contrast, philosophy appeals to those who have a more analytic frame of mind, where conceptual puzzles and arguments form the heart of matters. Somehow this pedagogical aspect had to be smoothly fused with the philosophical analysis. Not an easy task.

Any of these barriers would have delayed matters, but their combination has resulted in a very long delay between conception and production. Hopefully, this considerable period of gestation has resulted in a worthy introduction to the philosophy of computer science. Of course, I am still not satisfied with the result. Many of the chapters do little more than sketch the issues; many are just tasters. However, my hope is that the book will provide an overall conception, and a set of questions and puzzles, that will encourage more philosophers to engage with the material, and thereby inspire a deeper and more complete philosophical analysis of an area of knowledge that has transformed our contemporary world.

My greatest debt goes to Nicola Angius. Working with him over the past few years has helped me to shape my concept of the philosophy of computer science. Liesbeth de Mol and Giuseppe Primiero have, through HaPoC, provided many forums in which I have delivered keynote presentations that have been instrumental in developing many of the ideas of the present work. Their support and insights have been invaluable.

Discussions with Colin Phillips on Wittgenstein's philosophies of language and mathematics have greatly influenced my approach to the present subject. The notion that programming involves rule following is a theme that crops up throughout the book.

Discussions with the staff at my home university, especially Sam Steel and Chris Fox, have been instructive and encouraging.

Finally, without my friends and family I would not have started, let alone finished.

Colchester, April 2018 *Raymond Turner*

Contents

Part I
INTRODUCTION

This part of the book introduces its subject matter. The first chapter is a conceptual tour of computer science. It provides an introduction to the discipline that is intended to be accessible to philosophers who have little knowledge of mainstream computer science. The second chapter is intended for computer scientists who have little philosophical background; it introduces the main philosophical themes and concerns of the book.

Chapter 1
COMPUTER SCIENCE

The philosophy of computer science [233] is intended to stand to computer science as the philosophies of mathematics [121], biology [95], and technology [75] stand to their respective disciplines. These are instances of what Stuart Shapiro [215] calls *the philosophy of X disciplines*, and any philosophical investigation of such subjects must begin with some discussion of the discipline itself. In particular, it must aim to expose its subject matter, uncover its core activities, and draw out its distinctive features.

In this regard, computer science is a very broad church. It encompasses a large number of different activities that range from abstract mathematical topics to core engineering practices and scientific investigations [222]. The Association for Computing Machinery (ACM) is arguably the world's leading educational and scientific computing society. Its list of topics bears witness to the breadth of the subject. It includes hardware, computer systems organization, software and its engineering, networks, the theory of computation, the mathematics of computing, information systems, security and privacy, computing methodologies, and applied computing. Quite a list. In one way or another, such topics form the curriculum for most computer science departments, and might be taken to constitute its subject matter. However, by themselves lists are rarely enlightening. Is there something at the heart of all these seemingly disparate activities? Many authors have made suggestions.

1.1 Machines

Richard Hamming, a Turing Prize winner, makes computers and their use the heart of the discipline [102]:

> At the heart of computer science lies a technological device, the computing machine. Without the machine almost all of what we do would become idle speculation, hardly different from that of the notorious Scholastics of the Middle Ages.

© Springer-Verlag GmbH Germany, part of Springer Nature 2018
R. Turner, *Computational Artifacts*, https://doi.org/10.1007/978-3-662-55565-1_1

Although this does a disservice to the Scholastics, it does contain an important truth. It is undoubtedly the case that without computing machines the subject would be physically impotent. There would be no mobile phones, no laptops, no tablets, and no smart hair curlers or football boots. There would be no reasonably accurate weather forecasts, and no digital guidance systems. Indeed, there would be no direct physical applications of any kind. Of course, we could still construct algorithms, analyze them and run them by hand. That is, we could still do what ordinary people and mathematicians did before computers appeared. Nevertheless, without physical computers it is doubtful whether computer science would have emerged as a separate and distinctive discipline. Computers in all their forms provide the concrete mechanisms that gives computer science its physical potency.

On the other hand, while using computers may well be what most of these ACM activities have in common, their use alone does not characterize the discipline: hairdressers use them in their smart hair curlers, and hairdressing is not on the ACM list. Nor is football. While few would exclude computers and their applications as a defining part of computer science, many would resist the claim that they are its intellectual heart. One of the most famous computer scientists of the twentieth century, E. W. Dijkstra, is unequivocal in his denial of the centrality of computers [61]:

> I don't need to waste my time with a computer just because I am a computer scientist.

The following is less terse. It is a quote from a video course given at MIT based on a well-known introductory textbook on computer science, *Structure and Interpretation of Computer Programs* by Abelson et al [1]:

> [Computer science] is not really about computers – and it's not about computers in the same sense that physics is not really about particle accelerators, and biology is not about microscopes and Petri dishes ... and geometry isn't really about using surveying instruments. Now the reason that we think computer science is about computers is pretty much the same reason that the Egyptians thought geometry was about surveying instruments: when some field is just getting started and you don't really understand it very well, it's very easy to confuse the essence of what you're doing with the tools that you use.

These authors suggest that computers are the tools of computer science, and that we should not confuse the essence of the subject with its tools. However, this is only as persuasive as the analogy. If we replace physics with civil engineering and particle accelerators with roads and bridges, then it is somewhat less persuasive. While civil engineering is not about using roads and bridges, it is concerned with designing and constructing them. Likewise, computer science in one of its guises is concerned with designing and constructing machines. But I suspect some of the above authors intend to exclude even this as part of the core activity [56]:

> A confusion of even longer standing came from the fact that the unprepared included the electronic engineers that were supposed to design, build, and maintain the machines. The job was actually beyond the electronic technology of the day, and, as a result, the question of how to get and keep the physical equipment more or less in working condition became in the early days the all-overriding concern. As a result, the topic became – primarily in the USA – prematurely known as "computer science" which, actually is like referring to surgery

as "knife science" – and it was firmly implanted in people's minds that computing science is about machines and their peripheral equipment. Quod non.

Machines aside, computer science is undoubtedly concerned with the construction of many of its tools. For example, programming languages are an essential and central tool of the discipline, and it would be foolhardy to claim that computer science was not concerned with their design and construction. It clearly is. The tool metaphor is misleading.

1.2 Theoretical Computer Science

However, Dijkstra does clarify his stance. His main point seems to be that the intellectual core of the subject is largely independent of any physical implementation [57]:

> We now know that a programmable computer is no more and no less than an extremely handy device for realizing any conceivable mechanism without changing a single wire, and that the core challenge for computing science is hence a conceptual one, viz., what (abstract) mechanisms we can conceive without getting lost in the complexities of our own making.

Here he brings to the fore the abstract mechanisms of computing science that are independent of any particular physical representation. And for him it is the study of these abstract mechanisms or machines that constitutes the intellectual core of the discipline. Knuth expresses much the same thought in several of his famous quotes [132].

More generally, the core activity is something like the study of abstract machines, algorithms, and computational structures. These abstract notions are at the heart of the subject. After all, it is in this arena where the central intellectual ideas of the discipline are rehearsed. For instance, Turing machines are mathematical structures that have inspired generations of abstract machines. Similarly, algorithms can be followed and constructed, and computations induced with paper and pencil; no physical machines necessary. Indeed, the study and construction of algorithms precedes the advent of the computer by thousands of years. These abstract devices can be defined, designed, and explored without any recourse to physical devices.

This is a move in the direction of theoretical computer science and the theory of computation. Concerned as it is with the complexity and practical tractability of problems, computational complexity theory [91, 239] is a significant aspect of this. And so are formal language theory, type theory, type inference systems, semantic theories of programming languages, and domain theory. Undoubtedly, the mathematical investigation of such abstract computational mechanisms and structures is a significant part of computer science, and much of it is rapidly becoming a significant part of mainstream mathematics. However, taken as the core of the subject, this reduces computer science to the study of abstract mathematical structures, and by itself this is just as problematic as the purely physical orientation. Indeed, Hamming's central point is surely relevant here. For the discipline of computer science to exist as an independent enterprise, a discipline that is not a branch of mathematics, physical implementation is required. And this seems undeniable. Without computers there would be mathematics but no separate discipline of computer science. On the other hand, without theoretical computer science there would be no intellectual core; no mathematical foundations and no mathematical guidance.

1.3 Programs and Programming

Actually, the intent behind Dijkstra's remarks is broader. His abstract mechanisms have a practical intent: they are intended to be the mechanisms that are crucial to the activity of programming. Abstract mechanisms include the mechanisms and concepts built into contemporary programming languages. Indeed, the design of programming languages has generated many of them. These include the rich variety of type systems, nondeterminism and parallelism, the various paradigms of programming, etc. Another Turing Prize winner, Tony Hoare, puts matters as follows [117]:

> Having surveyed the relationships of computer science with other disciplines, it remains to answer the basic questions: What is the central core of the subject? What is it that distinguishes it from the separate subjects with which it is related? What is the linking thread which gathers these disparate branches into a single discipline? My answer to these questions is simple – it is the art of programming a computer. It is the art of designing efficient and elegant methods of getting a computer to solve problems, theoretical or practical, small or large, simple or complex. It is the art of translating this design into an effective and accurate computer program.

According to Hoare, the core activity of computer science concerns the design of correct, efficient, and elegant methods of getting a computer to solve problems. It is the activity that brings the abstract in contact with the physical. Indeed, according to Hoare it does more than this; it is what ties all the disparate aspects of the subject together. How so?

The activity of programming has inspired the invention of principles such as abstraction and modularity that help the programmer to design correct, transparent and elegant code. It involves thinking at different levels of abstraction, and decomposing problems into parts. It has generated the need for high-level programming languages, and without them contemporary programming would be impossible. Their construction has given rise to grammatical and type inference systems [187], techniques for their implementation, and principles of language design [168, 223]. They have also inspired mathematical approaches to their semantic definitions [213, 224]. Programming is also supported by the process of specification, and this has given birth to specification languages [127, 258]. Specification involves abstraction and decomposition: a complex problem specification must be decomposed into smaller ones, and specification languages must have mechanisms to support this [209, 258]. Programs have somehow to be shown to satisfy their specifications, and this has motivated the development of logical frameworks, type systems, proof systems, and verification tools. In addition, programmers must be concerned with the efficiency of their designs [91]. This has reinforced the study of algorithm complexity and enriched computational complexity theory. The development of software systems employs architectural description languages [49, 148] and software engineering methods [214]. These extend and generalize the core activity to cope with large, complex systems. Moreover, programming has influenced the design of actual machines: the design of physical machines that more closely resemble the underlying abstract machines of programming languages has formed a significant part of computer design. In particular, logic machines [244] were inspired by the demands of logic programming, and Landin's SECD machine [141] provided the abstract framework for the design of functional machines. The list of entailed and subsidiary activities and enterprises is long and encompasses much of the ACM list. These activities are subservient to the core; like baubles on a Christmas tree, they hang from it.

In this sense, programming unites the mathematical world of abstract machines and the world of physical ones. It brings together the two halves of computer science: the abstract mathematical world and the physical mechanistic one. Moreover, it has generated much of the research in the two halves of the subject: programming as an activity has enriched and motivated much of the research activity in both machine and hardware design and theoretical computer science. It is in this dual role of bridging and inspiring these disparate aspects that *programming a computer* is arguably the unifying topic of the discipline.

Despite this, many would still claim that it is the actual applications of programming, its constructed computational systems, that give the discipline its potency: it is these applications that have changed our lives. On reflection, this is really the significant point of Hamming's remarks. Applications include not just technological ones such as smart hair curlers and football boots, and systems that run nuclear power stations and guide missiles to their targets, but scientific ones such as those involved in computational biology and cognitive science. Indeed, the research agendas of many computer science departments, to say nothing of industrial and commercial establishments, are dominated by such topics.

However, no matter how useful and impressive these applications are, they have very specialized goals. Presumably, the goals of computational biology are biological and those of cognitive science psychological. And people who build flight control software are worried about planes crashing, and those who design smart hair curlers are concerned about hair that spontaneously straightens. In contrast, the Dijkstra-Hoare core of the subject does not have the goals of any particular application. While these applications may feed back into the core, and indirectly provide much of its content, on this view computer science is centrally concerned with the generic activity of programming a computer. An analogy may help. While pure mathematics is centered on *definition* and *proof*, the (Dijkstra-Hoare) focus of pure computer science is *specification* and *programming*.

1.4 Computational Thinking

There is what appears to be an even more abstract and encompassing interpretation of the core. The following is from Jeanette Wing's original manifesto on computational thinking [250]:

> Computational thinking is using abstraction and decomposition when attacking a large complex task or designing a large complex system. It is separation of concerns. It is choosing an appropriate representation for a problem or modeling the relevant aspects of a problem to make it tractable. It is using invariants to describe a system's behavior succinctly and declaratively. It is having the confidence we can safely use, modify, and influence a large complex system without understanding its every detail. It is modularizing something in anticipation of multiple users or prefetching and caching in anticipation of future use. Computational thinking is thinking in terms of prevention, protection, and recovery from worst-case scenarios through redundancy, damage containment, and error correction. It is calling gridlock deadlock and contracts interfaces. It is learning to avoid race conditions when synchronizing meetings with one another.

Broadly conceived, this seems to embrace the Dijkstra-Hoare characterization. These topics serve the central activity. Indeed, Hoare and Dijkstra themselves are responsible for the advocacy and development of many of them. Despite this, there is an apparent conflict. Wing goes on to say:

> Computer science is not computer programming. Thinking like a computer scientist means more than being able to program a computer. It requires thinking at multiple levels of abstraction ... Computational thinking is a way humans solve problems; it is not trying to get humans to think like computers.

Here Wing seems to be in direct conflict with Dijkstra and Hoare; she seems to be suggesting that programming a computer does not by itself involve thinking at multiple levels of abstraction, etc. In contrast, the Dijkstra-Hoare characterization embraces the idea that in constructing computer programs, computer scientists engage in activities such as problem decomposition and employ multiple levels of abstraction. Moreover, the use of high-level programming languages is aimed precisely at enabling programmers to be human problem solvers. Seen in their full glory, computer programming and software development involve all the topics in Wing's list. Still, I think her claim goes further. Regarding computational thinking, she says:

> This kind of thinking will be part of the skill set of not only other scientists but of everyone else. Ubiquitous computing is to today as computational thinking is to tomorrow. Ubiquitous computing was yesterday's dream that became today's reality; computational thinking is tomorrow's reality.

Although thinking like a computer scientist means employing programming methodology, it is to be applied in a more universal way: it should be applied to life. It is a fundamental skill that should stand alongside reading, arithmetic, and novel writing. There are many issues that arise from this apparently more universal enterprise. More specifically, does it require and entail an ontology of objects and concepts that are, in some sense, purely *computational*? Likewise, is there a form of knowledge that is neither scientific, mathematical, nor technological? And are there distinctive computational methods of obtaining that knowledge: is there a *computational methodology*? Whatever their merits, these questions are best addressed in terms of the Dijkstra-Hoare core activity. After all, it is inside the core where these notions were born, and computational thinking is the core activity applied more widely.

1.5 The Discipline

In their different ways, and for different reasons, each of these three areas of computer science (machines, theory, and programming) has a strong claim to prominence and centrality. While the activity of programming has this uniting function, without the other two there would be nothing to unite. Moreover, without actual computers there would not be an independent discipline of computer science. And, as a consequence, there would be no activity of programming, and none of the derivative activities. We would be left with theoretical computer science. Indeed, we would only have a fragment of this, since much of the mathematical development is parasitic on programming and programming languages. On the other hand, without programming, computers would be next to useless. And, without theoretical computer science, we would have no mathematical foundations for programming languages, no analysis of algorithms, and no support for the activity of programming and programming language design and implementation. Indeed, without Turing, Post, and Church, the first generation of theoretical computer scientists, we would have very little.

The discipline is a fusion of all three, where each seems to be ontologically and pragmatically dependent upon the others. Together they provide the unique character and substance of the subject. Indeed, these three aspects, and their relationships and dependencies, underlie many of the philosophical concerns of the discipline. They form the components of our notion of computational artifact.

Chapter 2
TOWARDS A PHILOSOPHY OF COMPUTER SCIENCE

In general, the philosophy of a discipline is characterized by Stewart Shapiro [215] as follows:

> For nearly every field of study, there is a branch of philosophy, called the philosophy of that field. Since the main purpose of a given field of study is to contribute to knowledge, the philosophy of X is, at least in part, a branch of epistemology. Its purpose is to provide an account of the goals, methodology, and subject matter of X.

Any philosophy of computer science [233] must describe and categorize the nature of its entities. In this regard, via their semantic definitions, the formal languages of the discipline somehow govern the subject's ontology. Consequently, some attention must be given to the nature of these semantic accounts, and how they fix or contribute to the kinds of entity dealt with. Furthermore, any statement of the goals of computer science will highlight its methods of reaching those goals, its methodology. This will bring to the fore its claims to knowledge, its epistemology. And, while addressing these issues, we must assess if computer science raises any distinctive philosophical concerns.

Exploring these issues will provide a philosophical analysis of the discipline structured along traditional philosophical lines.

2.1 Semantics

Computer science is a discipline dominated by languages. Aside from programming languages, the discipline has spawned languages for program and system specification, hardware description languages, database query languages, and web design and ontology languages. In particular, without programming languages the physical machines, advocated by some to be the heart of the discipline, would be idle devices. Programming brings the machines to life.

However, in themselves they are little more than string-generating systems, and while we may study them as such, without semantic content they are next to useless. On the face of it, symbolic

© Springer-Verlag GmbH Germany, part of Springer Nature 2018
R. Turner, *Computational Artifacts*, https://doi.org/10.1007/978-3-662-55565-1_2

programs are linguistic things fixed by the grammar of their containing language. For instance, the following simple program is written in the programming language Miranda [227].

```
ack  0  n  =  n+1
ack  (m+1)  0  =  ack  m  1
ack  (m+1)  (n+1)  =  ack  m  (ack  (m+1)  n)
```

You might be able to figure out what this program does if you understand recursion, or its name might give the game away. However, without some account of its semantics, you will have more problems with the following program:

```c
#include <stdio.h>
int  main()
{
int  n,  reversedInteger = 0,  remainder,  originalInteger;
printf("Enter  an  integer:  ");
scanf("%d",  &n);
originalInteger  =  n;
//  reversed  integer  is  stored  in  variable
while( n!=0 )
{remainder  =  n%10;
reversedInteger  =  reversedInteger*10 + remainder;
n /=  10;}
//  palindrome  if  orignalInteger  and  reversedInteger  are  equal
if  (originalInteger  ==  reversedInteger)
printf("%d is  a  palindrome.",  originalInteger);
else
printf("%d is  not  a  palindrome.",  originalInteger);
return  0;
}
```

Knowing a language involves more than knowing its syntax; knowing a language involves knowing what the constructs of the language do, and this is semantic knowledge. In general, without such knowledge it would not be possible to construct or grasp such programs. But how is this semantic content given to us; how is it expressed? What constraints or principles must any adequate semantic account satisfy? What are the theoretical and practical roles of semantic theory? Is natural language an adequate medium for the expression of semantic accounts? These are some of the questions that any conceptual investigation of programming language semantics must address.

There are also formal approaches to programming language semantics. These include operational, denotational, and game-theoretic accounts [68, 167, 168, 213, 224]. What are the relationships between these various approaches? Do they complement each other or are they competitors? Is one taken to be the most fundamental?

These and other questions parallel those of the semantic theories of natural language and the philosophy of language. In particular, we shall be concerned with the normative and compositional nature of semantic theories.

2.2 Ontology

A central topic in the philosophy of computer science concerns the ontological status of programs. While algorithms are generally taken to be mathematical objects, the nature of programs is less clear. And here semantic concerns are again central: their ontological status is closely allied to their semantic status. In particular, a semantic account of programming languages is taken to involve a machine of some kind. But what kind of machine? Is it abstract or concrete? If a physical machine is taken to fix the meaning, then semantically and ontologically, programs are primarily physical devices. On the other hand, if an abstract machine is employed, then programs are abstract in nature. However, the nature of programs is not so easily and cleanly settled: both the abstract and the physical devices seem to be involved. This much was highlighted in an early stage of the development of the philosophy of computer science by Moor [170]:

> It is important to remember that computer programs can be understood on the physical level as well as the symbolic level.

On the face of it, programs seem to be linguistic objects with an abstract interpretation together with a physical manifestation. In their abstract form, they appear to be abstract mathematical objects determined by the syntactic and semantic definitions of their containing language. However, in their physical guise they are technological constructions. Indeed, everywhere we turn in computer science we find the same pattern: a fusion of an abstract thing with a concrete one. Does this herald the introduction of a new kind of ontological entity? How are we to categorize and conceptualize the entities of computer science?

Our approach will be to treat programs as technical artifacts [139]. This will provide an existing and relevant philosophical framework for their analysis, and one that brings with it tools and concepts from the philosophy of technology [75].

2.3 Methodology

Methodology is the theoretical analysis of the methods and working principles of a discipline. Seen as design activities, software design and programming inherit some central philosophical concerns and questions from the philosophy of design [179]. What is a *well-designed* program or software design? What is a well-designed programming language? Which methods are employed for obtaining good designs, and how successful are they?

It almost goes without saying that correctness, the fact that a design meets its specification, is a necessary attribute of a well-designed program. Indeed, the computer scientist Tony Hoare [117] makes it the primary one:

> The most important property of a program is whether it accomplishes the intention of its user.

A second one, and one that contributes to correctness, is *simplicity*. This much is acknowledged by computer scientists of all varieties. But what is simplicity when applied to software designs and programs, and how is it achieved? Again, we shall attempt to place the discussion within the confines of an existing philosophical framework or area of investigation. In this regard, Baker [17] provides a starting point for a philosophical analysis of the notion of simplicity. However, the emphasis in [17] is on simplicity as applied to scientific theories, and little attention is paid to simplicity in design, and nothing to computational notions. Still, it provides a starting point and a framework for our analysis. However, we cannot take it for granted that the notions of simplicity for programs and scientific theories coincide. They do not. Part of our task will be to provide a clear exposition and analysis of the computational notion.

Many other methods and principles have been canvassed to support, enable, and encourage good system and program design. One of these is *modularization*. It has its origins in structured programming, and finds its modern incarnation in the central notions of object-oriented languages. Much of the philosophical discussion of these notions is located in the philosophies of mind and cognitive science [34, 73]. Our central task will be to provide a clear conceptual exposition and analysis of the computational notions of module, modularity, and modularization.

Another way of avoiding complexity in design involves *abstraction* [41]. While modularity tackles the problem of complexity from a horizontal perspective, abstraction works vertically: through levels of abstraction, it dissolves complexity by suppressing or hiding detail. Classical accounts of the notion [147] are not in fashion. We shall explore an approach to computational abstraction that has its origins in Frege [76], and is currently being explored in the philosophies of logic and mathematics [260].

Methods for program and system development also include formal methods [127, 258] that aim to yield correct programs and software. They seek to derive programs from their specifications, where the derivations employ simple transformation rules that preserve correctness; if correctly applied, they guarantee correctness. This is their virtue. On the other hand, there are thought to be serious practical limitations to the formal approach, and only a very small amount of contemporary software is generated in this way. Practical software must be efficiently produced: methods used in practice must work on a large scale, and the complexity of actual software attacks the very heart of the formal approaches. Generally, practice does not deliver full correctness. Mathematical proofs are usurped by testing and verification. But, as Dijkstra [57] points out,

> Program testing can be used to show the presence of bugs, but never to show their absence.

This conflict between correctness and practice is a serious methodological dispute at the heart of the discipline.

2.4 Epistemology

Practitioners of all forms possess *knowing how* knowledge, the kind of knowledge you have when you know how to do something. This is generally contrasted with the kind of knowledge you have when you know that something is true, i.e., propositional knowledge [197]. Presumably, programmers have a good deal of *knowing how* knowledge, much of which is associated with the concepts of the last section. In particular, knowing how to program large systems that meet their specifications requires knowledge of the techniques associated with modularization and abstraction.

However, there are also forms of *knowing that* knowledge. One such involves knowledge of the syntax and semantics of the programming language. Without such knowledge, the language could not be used. Another concerns Turing's analysis of computation. This is taken to provide a formal account of a central epistemological notion.

Still another is associated with the relationship between programs and their specifications. More specifically, issues arise concerning the form and content of correctness proofs and verification techniques: what kind of knowledge do they deliver? What kind of knowledge do programmers have when we have proven their programs to be correct? There are several challenges to the view that correctness proofs deliver mathematical knowledge. Indeed, they lead us away from the latter to a form of empirical agreement [70].

Actual correctness proofs are long, combinatorial and shallow. So, the question arises as to whether they are genuine mathematical proofs. This is the *mathematical* challenge. Because of their size, involving the checking of untold numbers of cases, these proofs are very often carried out with the aid of theorem-provers. This brings the computer, a physical device, into the arena of mathematical proofs. This is the *mechanical* challenge: such correctness proofs rely on a physical device, and so do not deliver mathematical knowledge. If this challenge succeeds, it would seem to follow that we do not have mathematical knowledge of programs. At best, we have some form of empirical knowledge of them. Indeed, in practice, because of their size and complexity, these proofs are rarely carried out, with or without the aid of theorem-provers: in practice, proofs are replaced with testing and verification. This is the *pragmatic* challenge. Judiciously designed tests are employed to examine whether the program meets its specification. But what kind of knowledge do these return? Once again, it is some kind of empirical knowledge, but what kind? Are these methods of testing and verification aimed at the same goals as those of mainstream science? This is the *scientific* challenge. These challenges are not entirely independent, and unraveling their nature and the connections between them will form the focus of our discussion.

There are two other possible challenges or puzzles that question what is actually achieved by correctness proofs. Even when carried out, they do not guarantee that the physical program, the one generated by the implementation, is correct. Even if we have correctness proofs for all the software involved in the implementation, we still have to deal with the abstract/physical interface. We still have no guarantee that the software will work. This is the *causal* challenge.

In turn, this generates the *trivialization* challenge, a conceptual puzzle first brought to our attention by Putnam [190]. What does the correctness of a physical device amount to? While correctness proofs relate two abstract descriptions, physical correctness connects an abstract description with a physical device. In regard to physical computations, and this applies to any physical device, Putnam points out that physical implementation seems only to demand an extensional correspondence between two domains: an abstract one and a physical one. And such correspondences are far too easy to come by [185]. In contrast, in practice, physical verification is a very complex and time-consuming affair.

Finally, there are questions about the nature of explanation in computer science. Asked to explain why it is that a program satisfies its specification, what kind of answer can or should a programmer give? This is obviously linked to the methods used to establish correctness.

2.5 Conclusion

As we said at the outset, structuring our investigation along these lines will give it a traditional philosophical appearance. However, as we proceed it will become clear that, as is the case with more traditional areas of the philosophy of X, these topics are deeply interrelated. In particular, semantics influences the ontology of computational artifacts, and methodology and epistemology are intimately linked. So, in the end, any such division is largely organizational. Still, it will facilitate a smooth presentation of the material centered on these traditional philosophical concerns.

At this point, it is worth noting a significant omission from the agenda: computational ethics. This is a large area with a great deal of existing literature. There is much to say about this topic from within the present perspective, but it will have to wait for another occasion, and maybe even another author. Otherwise, the book will not be finished for still more years.

There is also no account of the philosophical issues associated with application areas and, in particular, artificial intelligence. Once again there is a good deal of philosophical work, and so the same reason accounts for its absence.

Nor is the book meant to be exhaustive in terms of its coverage of computer science. We have concentrated on the topics that we believe raise the most significant philosophical concerns and issues.

Part II
ONTOLOGY

The *things* of computer science include desktops, laptops and tablets, programs, software, virtual machines, operating systems, compilers, interpreters, computer games, and much more. What kind of things are they? On the face of it, some of them are abstract and some of them are concrete. In particular, programs are often taken to be abstract entities, while laptops are physical. However, as James Moor pointed out in [170], it is not clear that computational entities are easily classified as one or the other. A more considered approach is required.

In this section of the book, devoted to the ontology of computer science, we provide an analysis of the things of computer science as *technical artifacts* [75, 139]. Seeing them in this way is a two-way street. On the one hand, it enables us to apply analytical tools and concepts from the philosophy of technology [75] to the technical artifacts of computer science. This provides an existing philosophical framework for their conceptual analysis. On the other hand, computational artifacts provide a rich supply of technical artifacts of varying degrees of logical sophistication. This brings a new level of complexity into the world of technical artifacts that should have some impact upon their general philosophical analysis. Moreover, this sophistication emerges from simple logic machines, and so it should provide some insight into the emergence of complex functionality.

Chapter 3
COMPUTATIONAL ARTIFACTS

Technical artifacts are taken to include all the common objects of everyday life, such as chairs, televisions, paper clips, telephones, smartphones and dog collars. They are material objects, the engineered things of world that have been intentionally produced by humans in order to fulfill a practical function. In this regard, they differ from naturally occurring entities such as stars and rivers [240]. A central part of the analytic philosophy of technology aims to understand their nature, design, and construction, and it is from this perspective that we approach the artifacts of computer science.

3.1 Function and Structure

One of the claims of the philosophy of technology concerns the ontological status of technical artifacts: they are said to have a dual nature.

> Technical artifacts are, at least prima facie, always physical objects, but they are also objects that have a certain function. Looked upon merely as physical objects they fit into the physical or material conception of the world. Looked upon as functional objects, however, they do not. The concept of function does not appear in physical description of the world; it rather belongs to the intentional conceptualization. Technical artifacts thus have a dual nature: they cannot exhaustively be described within the physical conceptualization, since this has no place for their functional features, nor can they be described exhaustively within the intentional conceptualization, since their functionality must be realized in an adequate physical structure. [135]

This insistence on the need for a purpose or function has led philosophers (eg. [135, 136, 138, 162, 216]) to argue that technical artifacts have a dual nature fixed by two sets of properties:

- functional properties
- structural properties.

© Springer-Verlag GmbH Germany, part of Springer Nature 2018
R. Turner, *Computational Artifacts*, https://doi.org/10.1007/978-3-662-55565-1_3

For example, the purpose of a clock is to tell us the time, but how it does it, whether it is analog, digital, or based upon the sun's rays, is part of its structure. Likewise, the purpose of a car is transportation, but how it provides it, whether it is petrol, diesel or electric, is part of its structure. Structural properties present the artifact as a physical object or device. In engineering practice, the functional and the structural properties are often not separated in the description of an artifact. Nor, in the standard documentation associated with such artifacts, is there any indication of which is which. Nevertheless, from a logical perspective they constitute two very different forms of knowledge: one describes the physical characteristics of the artifact, and the other asserts what it is intended to do. Both are necessary components of the notion of a technical artifact: without one, it won't work; without the other, we don't know what working amounts to.

Moreover, the relationships between function and structure are central to our grasp of these concepts. Getting from function to structure is a creative activity; it is the crucial design stage of engineering. It requires the skill and expertise of the design engineer. It is also not possible to extract function from structure, but not because it is practically difficult and involves creativity; it is not possible for an entirely different reason. While by testing and experimentation we may be able to figure out what a device actually does, this may not be what it was intended to do: the design engineer may have got it wrong. A functional description must provide a black-box description that expresses what it is intended to do. While the structure determines what it actually does, the function is supposed to tell us what it ought to do. That is, the notion of function, while having a propositional content, is intentional in nature.

3.2 Design and Manufacture

The output of the design process is a structural description of an artifact, and there might be many structural forms that fulfill a given function. For example, in the case of clocks, one might be digital and the other analog, whereas in the case of computers, one might be a laptop and the other a desktop. The design process may itself be divided into several stages, where more informative structural descriptions emerge at each stage. Even so, the final description does not have to be a *complete* structural description in the sense that all the physical properties need to be included. In one sense this is obvious. A structural description of a bicycle frame might demand that it be blue and made of carbon fiber. But there are many different grades of the latter. We might insist that it be made of intermediate-modulus fiber where less material is required to get the same stiffness, and therefore a lighter structure. The degree and amount of detail vary from structural description to structural description.

But what governs such detail in practice? In simple cases the structural description is direct and close to a complete description of the artifact. For example, to describe a clock as made of aluminum is to describe some of its physical characteristics. Other details might describe its internal mechanisms. However, some details may be left to the manufacturing process. This will be particularly pertinent to the kinds of artifacts found in computer science, where the distance between structural descriptions and physical devices is bridged by a complex process of implementation, and the input to this process may fall well short of a complete physical description. All that is required is that, relative to the implementation stage, there is sufficient detail to enable the production of the physical device; where detail is lacking, the implementation must supply it. So, what constitutes a structural description is relative to the state and sophistication of the implementation process. In some cases the latter is often automated, and involves no more input than the structural description itself. But this does not mean that there is no creative input in implementation. The latter is itself a complex technical artifact with a function and a structure. As such, it has associated design and implementation stages. The following schematic summarizes this view of technical artifacts:

$$\text{Function} \rightarrow \textit{Design} \rightarrow \text{Structure} \rightarrow \textit{Implementation} \rightarrow \text{Artifact}$$

This will be the perspective that we carry over to *computational artifacts* – the technical artifacts of computer science. Finally, notice that any manufacturing process generates instances of the design. The design of a Ford Focus does not correspond to one Ford Focus but to a *car kind*. In addition, the design might be parameterized for color or fuel type. So, when we use the term *artifact* we are actually referring to *artifact kinds*, where the actual physical objects are instances of these kinds.

3.3 Theories of Function

The duality thesis raises a crucial philosophical concern that is put in the following way by Kroes [136]:

> Exactly how the physical and intentional conceptualisations of our world are related remains a vexing problem to which the long history of the mind-body problem in philosophy testifies. This situation also affects our understanding of technical artefacts: a conceptual framework that combines the physical and intentional (functional) aspects of technical artefacts is still lacking.

How do we relate the functional and the intentional? There are two standard theories that aim to address this question: the *causal* theories and the *intentional* ones. Causal theories insist that actual physical capacities determine function. Cummins's theory of functional analysis is an influential example: *no material object, no physical capacity, no technical function* [47]. The underlying claim is that, without the physical thing and its actual properties, there can be no actual artifact. There is an obvious truth here: without the physical device, there is no artifact and so there is no mechanism, and so no function is materialized. In particular, without a physical program running on an actual physical machine there is no physical computation. But to conclude from this that the function is fixed by a physical device is to mistake a physical description for an intention.

To illustrate matters, consider a kettle. While it may boil water, it also, as a side effect, heats the material of the kettle itself. But what is it that determines that heat generation is a side effect and not part of the main function? More generally, how do we decide which physical properties form part of the function and which do not? On the simple-minded causal account of function, where the latter is fixed by the physical device itself, what the artifact actually does is its function, and so we cannot distinguish intended from accidental properties. If performance determines function, there is no notion of malfunction and correctness. The intentional aspect of function has been completely lost.

There has to be an intentional aspect, and the alternative theory embraces this: it puts agents and their intentions at the heart of any theory of function. The underlying assumption is that functions reflect the intentions of a designer. The theory begins with observations about everyday objects: for example, a television can serve as a doorstop or a toothbrush can serve as a cutlery cleaner. The fact that the same physical thing can serve many purposes has led some philosophers to suggest that the function of an artifact is dependent upon the decision of some agent – the one who decides it will be a toothbrush or a cutlery cleaner. It is agents who ascribe functions to artifacts. Good examples of this approach are [152, 201]:

> [t]he function of an artifact is derivative from the purpose of some agent in making or appropriating the object; it is conferred on the object by the desires and beliefs of an agent. No agent, no purpose, no function [152].

But the introduction of agents and intentions brings some well-known problems. Where in this perspective, is the function located? On the naive intentional view, it is located in the mental states of the agent. Unfortunately, in their crude form, such theories have difficulty in accounting for how they impose any constraints upon the actual thing that is the artifact:

> If functions are seen primarily as patterns of mental states, and exist, so to speak, in the heads of the designers and users of artifacts only, then it becomes somewhat mysterious how a function relates to the physical substrate in a particular artifact [138].

The problem with such mental theories is that we have lost all connection with the physical structure. As an agent, I can have all sorts of images in my head, but how do any such images connect with the structure of an artifact?

3.4 Computational Artifacts

How is the function expressed for computational artifacts, and how is the structure laid down? What is the relationship between the function and the structure, and between the structure and the actual physical device? How are we to deal with the problems associated with the causal/intentional views of function in the case of computational artifacts? Somehow, the intuitions behind both theories need to be incorporated into any sensible theory. We need the insight that without the physical device there would be no materialized function, and without the perspective of an agent there would be no intention. In particular, two different kinds of intentional agents need to be addressed: designers and users. In general, engineers hold that it is the designer and implementer that play the critical role in determining the functions of artifacts, and we shall follow this perspective.

Answering these questions for the artifacts of computer science will not only put a great deal of flesh on the notion of a technical artifact, but also bring out some of the central properties and idiosyncrasies of the computational family. One aspect involves the role of formal languages. At both the functional and the structural level, computational artifacts employ formal languages for the expression of their functional and structural properties. And there is a rich and varied collection of these languages. This is one of the central features of the computational family: their descriptions, both functional and structural, are given in specially designed formal languages. Very largely, this is a consequence of the need for clarity and precision in their characterizing descriptions. These claims and observations will take a more concrete form as we proceed.

In the rest of this part of the book, concerned with ontology, we shall unpack and illustrate these ideas by reference to the central technical artifact kinds of computer science: computers, programs and software systems.

Chapter 4
LOGIC MACHINES AS TECHNICAL ARTIFACTS

Languages and machines represent the two ends of the computational spectrum: the abstract and the physical. They come together at the digital interface, the very lowest level in the computational realm. Digital circuits are employed to store, communicate, and manipulate data. Flip-flops, counters, converters, and memory circuits are common examples. Their building blocks are called *gates*, the most central of which correspond to arithmetic and Boolean operations. These are simple *logic machines*, so named because they are intended to represent some form of numerical or Boolean operation. More complex machines are built from them by connecting and composing them in various ways, where the most general-purpose *register-transfer* logic machine is a computer.

In this chapter, we consider logic machines as technical artifacts. We begin with the basic building blocks: logic gates and adders. We shall then consider more complex systems, ending in the von Neumann computer considered as a technical artifact.

4.1 Function

A digital circuit is an electronic device; it is a physical thing. But its functional specification must say what it is intended to do. One way to achieve this is to say that it is intended, for example, to be an *AND* gate or a *half-adder*. While this describes its function, there is a caveat: such descriptions presume an understanding of the very notions of AND gate and half-adder. Without such, they cannot serve as inputs to the design process. We might refine matters a little by saying that the purpose of an AND gate is to compute the logical conjunction of two Boolean values. However, this only helps if we know what the terms *conjunction* and *Boolean values* mean. At some point, we need to employ terms in a language that is clear and unambiguous, and understood by the agents involved in the design and implementation processes.

© Springer-Verlag GmbH Germany, part of Springer Nature 2018
R. Turner, *Computational Artifacts*, https://doi.org/10.1007/978-3-662-55565-1_4

Furthermore, the *half-adder* notion is standard, and so its definition is common knowledge. However, in practice there will be a need to define and construct circuits that have never been constructed before, and so would not have been previously defined. In such circumstances, functional specifications must begin with a *definition* of the required circuit, and this must be expressed in a language that supports such definitions. It is the activity of specifying new circuits that requires a language that facilitates a precise expression of their function. This applies throughout computer science: as we move through the levels of abstraction, new higher-level languages are required.

In the case of logic machines we require a language, a method of representation, which enables us to specify their input/output behavior. For simple logic machines, we can do this in a straightforward way using *truth tables*. The tables in Figure 4.1 provide the definitions of the basic Boolean logic gates. Each takes two inputs and returns one output, and represents a variant of the standard Boolean connectives.

AND gate

Input A	Input B	Output
0	0	0
1	0	0
0	1	0
1	1	1

NAND gate

Input A	Input B	Output
0	0	1
1	0	1
0	1	1
1	1	0

OR gate

Input A	Input B	Output
0	0	0
1	0	1
0	1	1
1	1	1

NOR gate

Input A	Input B	Output
0	0	1
1	0	0
0	1	0
1	1	0

EX-OR gate

Input A	Input B	Output
0	0	0
1	0	1
0	1	1
1	1	0

EX-NOR gate

Input A	Input B	Output
0	0	1
1	0	0
0	1	0
1	1	1

Fig. 4.1 Definition of basic logic gates

Together with these Boolean gates, *adders* of various kinds form the building blocks for the arithmetic unit of a standard von Neumann computer. In the following example, a half-adder, there is more than one output:

A	B	S	C
0	0	0	0
0	1	1	0
1	0	1	0
1	1	0	1

This can be unpacked as the simultaneous definition of two operations, the sum and carry, where the former has outputs in the sum column and the other outputs in the carry one. This is already an extension to the simple truth tables of Boolean logic.

In turn, these form the building blocks for larger logic machines. These are more complex than the simple truth table definitions in that there may be two or more operations defined in one table. However, they are still precise black-box functional descriptions that provide no account of how the actual electronic devices are to be built; they do not describe how the computations are to be carried out. They are functional specifications, not structural ones, and they cannot be easily turned into the latter. They tell us what the actual physical device should do: *the what* not *the how.*

However, taken as functional specifications they graduate from being mere definitions to being specifications of actual digital devices. In this way, they are given governance over the construction of such devices; they provide the *correctness criteria* for such devices. To say how one builds a digital device that satisfies such functional demands is the role of the structural descriptions.

One final observation: as a specification language, beyond the specification of very simple devices, truth tables are not practical. Specifying complex circuits in terms of input/output works only for very simple devices. We shall see how this is overcome by using the same language, but with different interpretations, to express both function and structure.

4.2 Structure

The structural description of a technical artifact is the designer's view of the artifact. In the case of logic gates, the structure is expressed in a formal language of some form. In the case of simple logic circuits, we could employ a logical formula expressed in the language of propositional logic. For example, the following expresses a Boolean structure with three inputs (A,B,C).

$$(A \wedge B) \vee (B \wedge C).$$

The truth table informs us what to compute; the logical formula tells us how. From the present perspective, the difference between a truth table and a logical formula is function versus structure.

However, for practical and pedagogical reasons, the standard way of providing the structural description of logic machines employs digital circuit diagrams, where a digital circuit is constructed from logic gates.

Figure 4.2 is a common representation of the set of basic logic gates. Each logic gate implements a truth table function, with several possible outputs. In particular, each logical connective is represented by a different shape. Of course, theoretically we do not need all these as primitives, but it is practically advantageous.

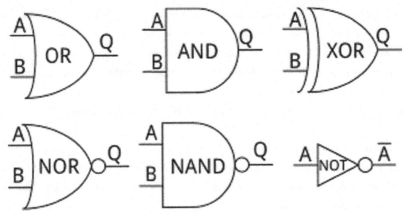

Fig. 4.2 Representation of basic logic gates

These gates may be combined in various ways to form more complex circuit descriptions. The circuit shown in Fig. 4.2 is an arithmetic logic unit. It is part of the central processing unit (CPU) of a computer that carries out arithmetical and logical operations on the operands in computer instructions. Some processors contain more than one arithmetic unit: they contain one for *fixed-point* operations and another for *floating-point* ones. Figure 4.3 displays a standard design, with separate logic and arithmetic units bolted together to allow communication between them.

This is a language of structural descriptions. It provides another example of how languages form the medium of representation in computer science. And, commonly, these languages enable the definition of artifacts of unbounded complexity, though clearly there is a practical limit as to what can be built.

How is the semantic definition of the language of digital circuit diagrams to be given? As one might expect, it is given by the truth tables themselves. For example, the semantic interpretation of the OR gate is given by the truth table for OR given above. The semantic account of more complex circuits is computed by composition. This is parallel to the way that complex propositional formulas have their truth tables computed. As a consequence, the language of truth tables serves two purposes: as semantic tools and as functional specifications. We shall have more to say about this shortly.

Actually, there are many languages available for describing the structure of such circuits. An alternative uses a hardware description language such as VHDL [13]. For example, the displayed VHDL code is for the half-adder. This is now recognizable as a program written in a programming language for digital circuits. While there are several largely equivalent ways of giving the structural

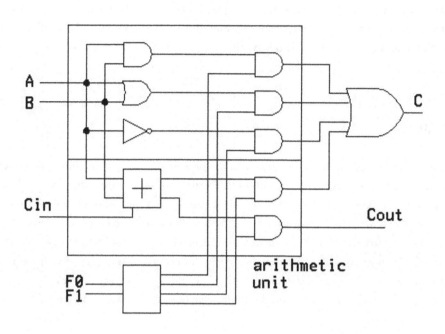

Fig. 4.3 A standard design of an arithmetic logic unit

```
VHDL CODE FOR half−adder :
ENTITY half_adder IS ―― half−adder
PORT(a,b:IN BIT; s,c :OUT BIT);
END half_adder;
ARCHITECTURE half_adder_beh OF half_adder IS
BEGIN
s <= a XOR b; ― Implements Sum for half−adder
c <= a AND b; ― Implements Carry for half−adder
END half_adder_beh;
```

description of such circuits, at some level they are all required to tell us how to build or construct the circuit.

It is characteristic of computer science that both functional and structural descriptions are given in formal languages that have been designed for this purpose. In the present case, these are the languages of truth tables and description languages for digital circuits. One tells us what to build and the other tells us how. However, here at least, the important point is not which language is used, but the fact that formal languages are employed, and have been designed for the purpose of providing both functional and structural descriptions. Designing languages is one of the central occupations of professional computer scientists, and their use marks one of the special features of computational artifacts. Most significantly, their logical complexity demands the use of precise formal languages.

4.3 Design

For technical artifacts, the top-level design step is the process of moving from function to structure. For example, given a specification of the half-adder in terms of inputs and outputs, the design task is to come up with a representation in terms of digital circuits. All this seems clear enough for simple cases, but computers are much more complex. As the complexity increases, the design of digital circuits becomes far from trivial. More complex logic machines are built from the basic building blocks, but exactly how they are put together is a complex programming task. In constructing such circuit diagrams, we are *programming* with logic gates. It is very low-level programming, but programming nevertheless. For example, the displayed VHDL code is part of the code for an arithmetic logic unit:

```
library IEEE;
use IEEE.STD_LOGIC_1164.ALL;
use IEEE.NUMERIC_STD.ALL;
entity alu is
port( Clk : in std_logic; ——clock signal
        A,B : in signed(3 downto 0); -input operands
        S : in unsigned(2 downto 0); -Operation
F : out signed(3 downto 0)  -output of ALU
        );
end alu;
```

The reader is not required to understand this code, but only to grasp the fact that it is an explicitly programmed version for part of the logic unit. Such examples prompt all kinds of conceptual concerns [179]. What is a good design? What methods are employed to obtain good designs? These questions will be posed for all of the computational artifacts we shall encounter. Whatever the other qualities a good design has, it must be correct: it must satisfy its functional specification. For digital circuits one of the other measures of a good design is given by the circuit's complexity. The lower the complexity, the lower the expense and error rate. Engineers use many methods to minimize logic functions in order to reduce a circuit's complexity. The most widely used simplification is a minimization algorithm such as the heuristic logic minimizer. Historically, binary decision diagrams, the Quine-McCluskey algorithm, truth tables, Karnaugh maps, and Boolean algebra have been used [7]. We shall discuss the issues that surround good design in a later chapter, but at all levels some notion of *simplicity*, the avoidance of complexity, is central.

4.4 Correctness

What is it for a structure, a digital circuit, to be *correct* relative to its functional specification? To illustrate matters, consider the AND gate Here the truth table provides the semantic interpretation of the digital circuit. Indeed, at the level of logic gates, the semantics of simple circuits (e.g. an AND gate) is given in terms of input/output tables. For simple Boolean circuits correctness amounts to the coincidence of its semantic definition with its functional definition. But matters are more involved for complex circuits. Just to establish that our half-adder circuit is correct relative to its functional specification, we would have to demonstrate that the two tables, the original one for the half-adder and the semantic one computed from the digital circuit by the composition of its components, are identical. In other words, correctness amounts to the claim that the two truth tables, the semantic one and the functional one, coincide. This is the correctness of the structural account relative to the functional description. And it is a formal or mathematical relationship.

However, notice that we cannot deduce or compute the function of the digital circuit from its semantic interpretation. This would be a category mistake that confuses a description of the circuit with a normative requirement for it. It would be to identify function with structure. The structure, the digital circuit, will be correct exactly when its semantic interpretation coincides with the truth table of its functional specification. As we said, while we may be able to figure out what the digital circuit actually does, in this case via its semantic interpretation, this may not be what it was intended to do. The latter is provided by the truth table that determines its intended function. The intentional aspect of functionality gives the independently given truth-table normative governance over the digital circuit.

4.5 Implementation

Given the structural description, the *manufacture* of physical digital circuits is a relatively straight-forward process. Physically the actual inputs might be an electrical flow or voltage that can, in turn, control more logic gates. For instance, in the diode of AND gate, when both the inputs are of same value, +5V, then the diodes are in OFF condition. As a result, no current flows through the resistor and there will not be any voltage drop across the resistor. Here the output will be +5V. Figure 4.4 provides the actual digital circuit for the AND gate.

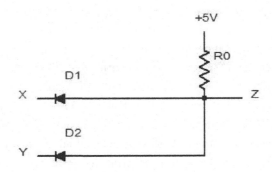

Fig. 4.4 Digital circuit for AND gate

It is here that the symbolic world ends. Digital circuits employ a very direct relationship between structural definition and artifact: the electronic circuit is fabricated from the digital structure. Fabrication is itself a technical artifact, whose function is to transform digital diagrams into their corresponding electronic ones.

4.6 Logic Machines as Technical Artifacts

In summary, we have the following picture of simple logic machines as technical artifacts.

Truth table → *Design* → Digital circuit → *Fabrication* → Electronic circuit

In the more complex cases, more sophisticated techniques of functional specification are employed. These machines constitute very simple and clear examples of technical artifacts where the structural description, the designer's view of the artifact, and the output of the implementation process are very close. While this will not be the case for all computational artifacts, they clearly illustrate their rather special nature, where functional and structural descriptions take on an abstract, even mathematical, form, and where design is a programming activity.

4.7 The Abstract-Concrete Interface

In the case of digital circuits, there is an intended relationship between the physical circuit (the actual artifact) and the structural description given by a digital circuit diagram. In this case the desired relationship is expressed as structural isomorphism. But, in itself, this will not guarantee any form of behavioral correctness i.e., we require the physical circuit to behave in a way that is in agreement with the digital circuit. This amounts to the demand that the physical device behaves as determined by the abstract input/output behavior of the abstract digital circuit.

The notion of agreement that we are looking for is the correspondence between the truth table that characterizes the digital circuit and the extensional physical behavior of the electronic device: if we run all combinations for the electronic device and make a table of its input/output behavior, this must be in one-to-one correspondence with the truth table that provides the semantics of the digital circuit. It is this that links the abstract and the physical. Of course, in practice this is impossible for all but the very simplest devices. But even when this is completed there is no mathematical guarantee of correctness. Unlike the correctness between the functional and structural specifications, this is an empirical notion of correctness that is to be tested by physical, not mathematical means. This is the abstract-concrete interface. The functional specification is a mathematical object – a truth table description of the input/output behavior of the circuit. Indeed, together with its truth table semantics, the structural description is also a formal mathematical device whose input/output behavior we can reason about independently of any physical realization in an electronic or any other physical medium. In contrast, an electronic device is a concrete physical thing. There are conceptual issues here having to do with the problem of characterizing computation and correctness for physical computations [185]. These concern the so called *simple mapping account of computation*, a topic for a later chapter.

4.8 The von Neumann Computer as Artifact

We have been building up to the von Neumann architecture (Figure 4.5). This is the computer architecture described in the first draft of a report on the EDVAC. It describes a design architecture for an electronic digital computer with parts consisting of a processing unit containing an arithmetic logic unit and processor registers, a control unit containing an instruction register and a program counter, a memory to store both data and instructions, external mass storage, and input and output mechanisms.

This is a higher-level piece of notation, which is unpacked by the structural description of each of the components, with a full digital diagram for the whole architecture.

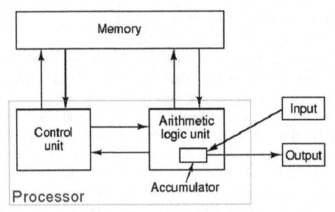

Fig. 4.5 The von Neumann architecture

4.9 High-Level Functional Notation

Although this complex structure is theoretically specifiable as a truth table, it is not practically feasible. However, figure 4.5 can be taken as a functional specification of a computer, with each of the components specified as truth tables. If we enrich the language of functional specifications to include such diagrammatic notation, then we have a more realistic language of functional specifications. As a functional description, it tells us that a computer must have a memory, a control unit, an arithmetic unit, and input/output devices. These have information flow between them, indicated by the arrows. Information is to be passed from the input to the arithmetic logic unit and, after processing, using any memory devices, to the output. This is a pictorial language for putting components together. As a functional definition of the stored program computer, its components are functional specifications of each of the components. So the reference to the arithmetic logic unit is a reference to the truth table definitions of the unit.

However, the same diagram may be interpreted structurally, in which case it is unpacked by the structural description of each of the components, resulting in a full digital diagram for the whole architecture. While there is ambiguity in these diagrams about how they are to be interpreted, functionally or structurally, the two interpretations have distinctive logical roles. Indeed, this illustrates how the difference may only be located in how we interpret the role of the diagram, not in the diagram itself. It is the intention behind them that is different and distinctive.

4.10 Conclusion

Computational artifacts are the technical artifacts of computer science. As such they have a function, a structure, and a physical existence. The functional and structural descriptions are given in formal languages designed for the task. Logic machines provide clear examples. In simple cases, the structural descriptions and the actual artifacts are structurally close together, and the latter are directly fabricated from the structural descriptions. For more complex computational artifacts this will not be so. In particular, it is not so for high-level programs which provide even more sophisticated examples of technical artifacts. From now on, the levels of abstraction that underpin computer science come to the fore.

Chapter 5
THE ONTOLOGY OF PROGRAMS

The ontological standing of programs has been the subject of some debate in the relatively small amount of philosophical literature that concerns mainstream computer science. In particular, much of this literature would have it that programs have both a symbolic or abstract representation and a physical manifestation. The earliest example of this is the following [170] by Jim Moor:

> It is important to remember that computer programs can be understood on the physical level as well as the symbolic level. The programming of early digital computers was commonly done by plugging in wires and throwing switches. Some analogue computers are still programmed in this way. The resulting programs are clearly as physical and as much a part of the computer system as any other part. Today digital machines usually store a program internally to speed up the execution of the program. A program in such a form is certainly physical and part of the computer system.

Seemingly, programs may be understood as symbolic entities that have a physical realization. A later expression of much the same observation, but put in terms of *text* and *machine*, was made by Tim Colburn [42]:

> Software seems to be at once both textual and machine-like. After all, when one looks at a printout of a program one sees a lot of statements written in a formal language. But when one holds the same program on a floppy disk in one's hand, one feels the weight of a piece of a machine.

The following quote by Nurbay Irmak is of more recent origin, and explicitly asserts the duality thesis:

> Many philosophers and computer scientists share the intuition that software has a dual nature. It appears that software is both an algorithm, a set of instructions, and a concrete object or a physical causal process [125].

Notice that a shift occurs here. The terms *set of instructions* and *algorithm* are both used to refer to the nonphysical guise of the program, and are contrasted with the physical realization. This suggests that the nonphysical guise is formal or mathematical in nature. Bipin Indurkhya in [124] is even more explicit when he suggests that the symbolic program be considered as a mathematical object:

© Springer-Verlag GmbH Germany, part of Springer Nature 2018
R. Turner, *Computational Artifacts*, https://doi.org/10.1007/978-3-662-55565-1_5

Software is a rather unique entity. On the one hand it can be considered a mathematical object – its component parts and operations of construction are rigorously defined, and the output result of a piece of software can be predicted precisely, at least in principle. On the other hand, it is also an empirical object – a piece of software executing on a machine is a physical object that can, as most of us have experienced on many occasions, produce unexpected and unforeseen behavior.

This duality has also been observed by lawyers concerned with the legal status of programs:

As I see it, the phrase "computer program" has two quite distinct primary meanings. One meaning refers to the text of the program that is made up of signs and symbols which may be fixed in some tangible medium of expression or transmitted as a stream of binary digits or other signals from one installation to another. The other meaning refers to the process that takes place inside a computer when the steps of the program are carried out [129].

All seem to agree that programs have abstract and physical manifestations. Indeed, somehow they have both. This raises a question about how one thing can exhibit such different guises. This apparent ontological dilemma Tim Colburn [42] expresses as follows:

While computer scientists often enthusiastically embrace this duality, a metaphysician will view it as a puzzle to be explained. How can something, namely a computer program, be at once concrete and abstract?

Whether there is a serious metaphysical issue here remains to be seen. There certainly would be if a single thing was somehow both concrete and abstract, whatever that might mean. However, even if there are two distinct things, they do appear to be tied together in some very intimate way, and this requires some investigation.

How can these two conceptions of the nature of computer programs, the abstract and the physical, be unified to provide a coherent ontology of programs? How are the above references to some kind of abstract mathematical entity to be understood, i.e., how exactly are programs to be taken as mathematical entities? Finally, notice that none of these quotes mentions the function or purpose of a program. But can one provide an ontological analysis of programs without some mention of their intended function?

Both programs and technical artifacts are said to have a dual nature. However, it is unclear how these dualities map onto each other. How are programs conceived as technical artifacts?

5.1 The Functional Specification of Programs

A functional description is intended to be a black-box description that tells us nothing about the structure of the device. It must inform us what the device is supposed to do – not what it actually does. In computer science, such functional descriptions of programs are called program specifications. A familiar and simple example concerns the greatest common divisor (GCD) of two numbers:

- A number z is the *greatest common divisor* of two numbers x, y if z divides both and is the biggest one that does.

This is a definition of the GCD: it does not tell us how to find the greatest common divisor; it only tells us what it is. In the language of technical artifacts, this is the *function* of a potential program. The greatest common divisor relation

$$Gcd(x, y, z),$$

is more formally defined by the following predicate calculus expression:

$$Divides(z, x) \land Divides(z, y) \land \forall w \bullet Divides(w, x) \land Divides(w, y) \to w \leq z.$$

In a similar manner, the following is the definition of a sorted list. It insists that the relationship between the input and output lists is such that the output is a sorted permutation of the input.

- A list of numbers x is a *sorted variant* of a list y if x is a permutation of y and y is sorted in ascending order.

Again, the corresponding predicate calculus expression more formally defines the notion (e.g., for lists of numbers):

$$Sort(x, y) \triangleq Perm(x, y) \land Sorted(y)$$

We could also provide formal versions of *Perm* and *Sorted*, but the general idea should be clear enough. Again, this tells us what sorting is, not how to sort: it clearly and precisely states the intended function of the required program.

In contrast to the general case of technical artifacts, these are very precise functional specifications that are expressed mathematically. But then programs are very precise things. While we may well know the concept of GCD or sorting and, therefore, may not require such precise and explicit statements, this will not be generally so. Indeed, because of the need for precision, and the complexity of software systems, computer scientists have designed specification languages alongside their programming cousins. And like their cousins, they have a rich type structure and structuring mechanisms [127]. We shall consider these in a later chapter where we consider the languages of computer science. For the moment, the central point concerns the nature of such specifications. Whether they are formal or informal, in themselves they are stipulative definitions [98]. At the level of abstraction at which they are formulated, definitions do not tell us how to build or construct the intended artifact. For example, the definition of a GCD determines an abstract notion that, in its natural setting, is part of number theory.

However, in computer science, such definitions are put to work as functional specifications of actual programs. When taken as specifications they are given a new occupation; they are given governance over any purported program. As such, definitions are taken to point beyond themselves to the construction of an artifact. When presented with the symbolic program, we might be able to guess or work out what it actually does. However, as we have previously emphasized, this may not be its function; the function is what it is intended to do, not what it actually does. While these two may coincide, indeed we aim to make them so, they may not. The program may not meet its specification; it may malfunction. Such specifications are normative in the sense that they fix the criterion of correctness and malfunction for any proposed program.

Should we employ algorithms as specifications? To see why this is undesirable, consider the following well-known algorithm for computing the GCD:

EUCLIDEAN ALGORITHM
1. Input two integers x,y.
2. If x<y, exchange x and y.
3. Divide x by y and get the remainder, r.
4. If r=0, report y as the GCD of x and y.
5. Otherwise replace x by y and replace y by r.
6. Return to the previous step.

This is based on the principle that the greatest common divisor of two numbers does not change if the larger number is replaced by its difference with the smaller one. Given that it is written in English, presumably in a precise and unambiguous way, we can reason about it in an abstract way. For example, we may formally argue that it terminates, and this is based on the observation that the remainder gets smaller on every iteration.

Can it be taken to be a functional specification? The fact that it contains a means of actually computing the GCD counts against it. The function is intended to tell us what is to be done at a suitable level of abstraction, and algorithms make too many concessions to the *how*. However, that is not the heart of the matter. The conceptual content of a specification must introduce or define what the artifacts are supposed to do. In this case it must be a *definition* of the GCD, and the Euclidean algorithm is not generally taken to be such a definition. Nor is Stein's algorithm that is given below.

These are different ways of computing the same thing. But what is that thing? Without a more abstract notion of the GCD, we could not say that these two were equivalent, i.e., compute the same thing. This definitional nature is a characteristic feature of computational functions. The underlying mathematical content of a specification, the content abstracted away from its intentional role, is definitional.

STEIN'S ALGORITHM

1. If either of the two values is zero,
the result of the algorithm is the other value.
2. If the two integer values are equal, the GCD
is this value.
3. If both of the values are even numbers,
we divide both values by two.
Find the GCD of the two new values,
and multiply the result by 2.
4. If only one of the values is even,
we divide the even value by two, and
recalculate the GCD.
5. If both of the values are odd,
the smaller value is subtracted from the larger,
and the result is used with the smaller value
to calculate the GCD.

Of course, matters are not so simply decided. Definitions come in all shapes and sizes. For example, consider the following definition of the factorial function:

$F(1)=1$
$F(n+1)=(n+1)*F(n)$

Is this a definition or an algorithm? It is easy to see it as both. On the one hand, it tells us what the factorial function is; it defines it. On the other hand, it also appears to provide a means of computing it: to compute factorial 4, we compute factorial 3, etc. However, this is not part of its definitional guise; it is part of an implicit algorithm that we are imposing upon matters. Considered as a definition we are only told the relationship between factorial $n+1$ and factorial n. We are not given any algorithmic information.

However, while there is a clear mathematical distinction between the definition of the GCD and any algorithm that computes it, in general the choice of the language of functional descriptions is methodological: we wish to minimize *how to* detail. What is not negotiable is the intentional role of functions: the defining distinction between functional and structural descriptions is intentional, and concerns what we take to have governance over what.

5.2 Structure

Are algorithms more likely candidates for structural descriptions? If they contain too much *how to* detail to play the functional role, does that not make them suitable candidates to play the structural one? Consider Euclid's algorithm. While it is written in vernacular, in English, it clearly supplies a method for computing the GCD. Moreover, many programmers will use an algorithm as the first step in the design of an implementable program. For example, the following program, which is written in the WHILE programming language [48], encodes the above algorithm for computing the GCD.

```
var       x,y,r: nat
begin
read  x,y;
          r:=x mod y;
          while  r  notequal  0  do
                 x:=y;  y:=r;  r:=x  mod  y
          od;  write  y
end
```

Similarly, the following encodes the binary method that is based on Stein's algorithm.

```
Input:  a,  b  positive  integers
Output:  g  and  d  such  that  g  is  odd  and  gcd(a,  b)  =  g  x  2d
    d  :=  0
    while  a  and  b  are  both  even  do
        a  :=  a/2
        b  :=  b/2
d  :=  d + 1
    while  a  notequal  b  do
          if  a  is  even  then  a  :=  a/2
          else  if  b  is  even  then  b  :=  b/2
          else  if  a > b  then  a  :=  (a–b)/2
          else  b  :=  (b–a)/2
g  :=  a
Output  g,d
```

So are such algorithms suitable vehicles for structural descriptions? One might insist that algorithms are natural language programs, i.e., special cases of programs written in natural language.

But there are obvious objections to this. For one thing, natural language programs have no direct implementation, and, ideally, structural descriptions must provide the input for a manufacturing process that is as mechanical as possible. For this we require programs that are expressed in an implemented language.

A second practical reason why structural descriptions need to be expressed in actual programming languages concerns the nature of programming as a design activity. Programming is a technical activity that is poorly served by natural language. The modern programmer designs in terms of the types and operations of her adopted high-level language. For example, the Haskell [226] programmer employs polymorphic function definitions. She programs to maximize generality: a polymorphic type may contain type variables, and these may be instantiated for any type of the language. The Java [200] programmer thinks in terms of classes and objects: she packages information in bundles that include the operations of the class. Programming is a technical endeavor that requires technical notions not available in natural language, which is impoverished if seen as a programming language. In practice, for pragmatic reasons, symbolic programs expressed in a *high-level* language provide the best candidates for the structural descriptions of programs.

5.3 Implementation

For computational artifacts, the physical device is the result of some process of implementation. In the case of programs, implementation is a mechanism that, given a symbolic program as input, returns a physical process. Even so, the symbolic program is not a complete structural description of the artifact; in practice, it results only in a structural account that is sufficient for the implementation process to kick in. By suggesting this analysis, we are allowing long-distance relationships between the structural description and the actual physical device. In doing so, are we being too generous with the term *structure*? Are we stretching it too much?

In practice, a structural description will not spell out all the physical details; many of these will be built into the implementation process itself. For example, in computer-aided manufacturing (CAM), software and machinery facilitate and automate manufacturing processes. Many decisions about the physical artifact have already been built into the manufacturing process; the designer designs up to the start of the manufacturing process. Consequently, a physical description does not have to be, and rarely is, a direct one. To illustrate matters, consider a punch card (Figure 5.1)

This physical card determines a physical program in the following sense: given a machine that inputs such cards, a physical program results. Together with the underlying mechanism that interprets the cards, the card determines the physical program. It describes the program in sufficient detail to enable the manufacture of the physical artifact. Here the structure of the technical arti-

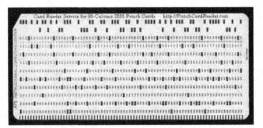

Fig. 5.1 A punch card

fact acts as an intermediary between its function and the physical device that is generated by the mechanism that interprets the card. The structural description, while it is the output of the design process, it is also the input to the manufacturing one.

In the more general case of a high-level program, layers of implementation stand in between the symbolic program and the physical device. Programming language implementations are complex pieces of software and hardware. In particular, they usually involve compilation, i.e., a translation from programs in a host language into programs in a target language. Eventually, the compilation process results in assembly language instructions that refer directly to a machine's architecture, including a hardware register. In this sense, the symbolic program, via implementation, reaches down to the physical one. Although the physical program is conceptually a long way from the high-level symbolic one, the implementation physically links the two. In this sense, the symbolic program acts as a structural description: via the implementation, this description is sufficient to generate the physical program, the thing that actually runs on a concrete machine. This generalizes the notion of physical description to include the case where the implementation process acts on the structural description to fix the artifact.

So, complex implementations are the means of seeing symbolic programs as structural descriptions of physical programs, their medium of execution. This provides a generalization of the notion of structure for logic machines, where the structure and the actual artifact are almost structurally identical. For programs in high-level languages, there is another creative step that involves the construction of the implementation. However, this is the responsibility of the implementer rather than the programmer. Moreover, as we shall see later, the implementation is itself a computational artifact.

5.4 The Symbolic and the Physical

This brings us to the basic ontological dilemma mentioned in the introduction to this chapter. How does this enable us to conceptualize the relationship between the two guises of programs? This is a more complex case than that of the digital circuits considered in Chapter 4. Colburn [42] suggests that software is a *concrete abstraction* that has a medium of description (the text, the abstraction) and a medium of execution (e.g., a concrete implementation in semiconductors). He suggests that there is only an appearance of synchronicity, in that a master craftsman (a computer scientist) has created two programs of such perfection that they always do the same thing. He refers to a notion of *harmony* between the symbolic program and the physical one:

> But the nature of software's duality can be illuminated by understanding the distinction between software's medium of description and its medium of execution. But the pre-established harmony thesis is well suited for explaining the high correlation between computational processes described abstractly in formal language and machine processes bouncing electrons around in a semiconducting medium. For, of course, it is not necessary to appeal to God in accounting for this correlation; it has been deliberately brought about through years of co-operative design of both hardware processors and language translators. An analogy used to describe parallelism in the mind/body problem imagines two clocks set and wound by God to tick in perfect synchrony forever. For the abstract/concrete problem we can replace God by the programmer who, on the one hand, by his casting of an algorithm in program text, describes a world of multiplying matrices, or resizing windows, or even processor registers; but on the other hand, by his act of typing, compiling, assembling, and link-loading, he causes a sequence of physical state changes that electronically mirrors his abstract world. The parallel nature of the abstract and the concrete is a defining characteristic of the digital age.

This notion of harmony obviously captures the idea that the two programs are in some agreement, and the master craftsman metaphor is instructive. However, it is a metaphor. Fortunately, we are able to be a little more explicit. In the case of programs, the symbolic and physical programs, *the medium of execution*, will be in *harmony* if the implementation is *correct*. This captures the informal idea that hardware processors and language translators are in agreement. But it is not about how they came to agree, but a statement of what that agreement amounts to. This characterization of harmony between the symbolic and the physical guises of programs unites them via an implementation that maps one to the other. And it does this indirectly. The statement that two programs are in *harmony* means that the physical program is a correct implementation of the symbolic one. We shall have more to say about this when we discuss both semantics and implementation in more detail.

However, there is some hidden complexity here. In the quote from [42], the symbolic form of the program is referred to as *abstract* whereas the quote from [125] employs the term *algorithm*. Indurkhya [124] comes clean and seems to classify the abstract form as mathematical. How do these abstract/mathematical guises fit into the symbolic/physical duality? The simple answer is through the semantics of the host programming language. Presumably, it is via the semantics that the programmer is able to design the program from the specification, and it is via the semantics that the programmer is able to explain why and justify the claim that the program meets the specification. Via the semantics of their containing language, symbolic programs are abstract mathematical objects. We shall put more flesh on this claim later. However, programs are not purely mathematical entities. As they are technical artifacts, the mathematical nature of the symbolic entity has to be physically manifested. In this sense, the concept of a technical artifact provides the unifying notion for the symbolic and physical guises of programs.

5.5 Programs as Technical Artifacts

Our original duo of symbolic program and physical manifestation has been replaced by the trinity of *specification*, *structure*, and *artifact*:

Specification → *Design* → Symbolic program → *Implementation* → Physical process.

Specification provides the function, a symbolic program is taken as the structural description, and the physical process is generated by the implementation. This provides a more robust and complete conceptualization of the ontology of programs. We shall call these ontological bundles *program artifacts*. These entities are complex packages of abstract concepts and physical devices or computations that are tied together via their correctness criteria.

Chapter 6
SOFTWARE SYSTEMS AS TECHNICAL ARTIFACTS

Our institutions and organizational structures, whether in government, commerce, industry, or education, are underpinned and controlled by software systems. So are our laptops, televisions, cars, and mobile phones. In this chapter we present a conceptualization of them as technical artifacts.

Whichever way they are put together, and whatever form their interactions take, the process of software development has different parts or components. Requirements analysis, system design, implementation, and verification are the central ones. The requirements stage gathers information about the required functionality of the project. The design of the system yields its overall structure, presumably in line with the functional demands delivered by the requirements analysis. The implementation outputs the coding of the system components, while the process of verification seeks to guarantee that the individual modules, and the overall structure, satisfy the functional demands.

While these are the basic stages of software development, different methods employ them in different ways [214]. More precisely, they differ in terms of the regimentation, rigidity and intensity of their employment. However, in this chapter our focus is not on these different methods of software development; that is a topic for the methodology part of the book. Here we aim only to put enough flesh on them to conceptualize software systems as technical artifacts.

6.1 Requirements

Any type of project, large or small, will involve some form of requirements gathering. Even the most "agile" developer must employ some initial account of what is to be constructed. The depth of analysis, the number of revisions and the techniques involved may differ from project to project, and from method to method, but some requirements there must be [112].

To illustrate matters, we consider the following scenario. The following is a very rudimentary set of requirements that might form the initial demands of an agile-style project [112]:

© Springer-Verlag GmbH Germany, part of Springer Nature 2018
R. Turner, *Computational Artifacts*, https://doi.org/10.1007/978-3-662-55565-1_6

A town requires a new library system. It must support the cataloging of new books, and the borrowing and returning of books by readers. It must have a system for acquiring books from vendors, and it must support all the normal security and financial arrangements concerning the borrowing and returning of books.

These requirements form a rather crude set of functional demands for a software system. The various software development methods would enrich, modify or change it at various stages in the overall process. At this top level of abstraction they are expressed in English vernacular.

6.2 Structure and Design

The term "software design" may refer broadly to the whole process of constructing a software system, and this is consistent with the fact that design occurs throughout. But we are concerned here with the stage following the requirements specification, the stage where the requirements specification is massaged into a system design. In the case of our library system the design process might result in a UML class diagram as shown in Figure 6.1.

Here the structural description of the system consists of several interconnected UML classes. Specifically, it describes five classes that correspond to the library itself, the various library items such as books and DVDs, and readers or patrons of the library. Each class is characterized by a set of attributes or properties, and a set of methods or operations that can act on the objects of the class. For instance, the library item class has the various attributes of a library item. These include its call number, title, cost and status, together with any operations one can employ such as adding a new item and finding its location. Of course, there has been much fleshing out of structure - but that is part of what a structural description for a software system must do.

Of course, we could have used any specification language with the ability to articulate complex system specifications. For example, we could have employed a Z specification of the notion of a book that would have the following form:

$$Book = [x : Isbn, y : Author, year : Year, ...]$$

While the specification mechanisms of these languages differ considerably in terms of how such information is represented and encoded, they each aim to provide a design of the whole system with each of its components represented by a module, class or schema, or in whatever representational mechanisms the language provides. Again, we shall have more to say about this when we discuss the languages of computer science. For now, we note that the outcome of the design stage is a structural description of the whole system. Notice how we have moved everything up one level from simple programs: what might have acted as a functional description for programs now forms part of the structural description of a whole system.

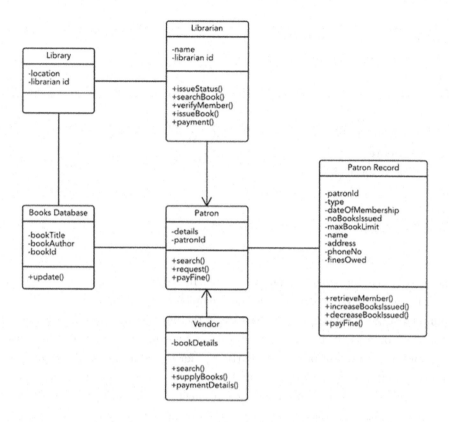

Fig. 6.1

6.3 Implementation

Next, the structure of the system must be filled out by the programming of the individual modules. Since programming is itself a design activity, this is also a design stage. The UML diagram is acting here as a specification of the program. In this case it is a minimal specification. For instance, it only demands that the book should be issued. We have to rely on our natural language understanding of what it is to issue a book. We could, of course, use more precise and formal specifications using a specification language such as Z or VDM. However, whatever notion of specification is employed, it generates a new level of design activity in that each individual method has to be implemented. For example, the following Java class might be taken as an implementation of the operation of issuing a book. The reader unfamiliar with Java need only grasp that implementation results in the following code that satisfies the UML specification:

```
class Issue : public Event {
  Issue
  public:
  Issue( person* _p, book* _b ) { _b = b; _p = p; }
  void operator ()() {
        require( _b && _p ); // _b and _p existDVDs
        _b->Issue(p);
        _p->allocate(b);
        }
  private:
  person* _p; book* _b;
}
```

But this is only the first stage of implementation. In order to maintain the notion that function and structure fix physical devices, we must employ the implementation of the containing programming language. This supplies the passage to the physical processes. Notice how the term *implementation* is overloaded. It is used in software systems for the step from UML specification to program, and, in the case of programs as artifacts, for the step from symbolic program to the physical process. In this perspective implementation has two stages: the first outputs a collection of symbolic programs, and in the second these are implemented as physical processes. Thus what was a structure in the software artifact now becomes the function one level of abstraction down. This perspective maintains the physical nature of actual artifacts. It allows us to distinguish between the abstract notions that inhabit functional and structural descriptions and the physical nature of actual artifacts. Consequently, it allows us to employ the conceptual framework of technical artifacts as it is. By incorporating levels of abstraction in the implementation we maintain that computational artifacts are ultimately physical devices. What prevents computer science from being a purely abstract discipline, what gives it its distinctive character, is the blending of the abstract and the physical in this notion of a computational artifact. The latter are physical devices that perform physical computations; without this insistence, computer science is in danger of falling back into mathematics.

6.4 Software Systems as Artifacts

There is a natural way of seeing matters that is conceptually clean and employs our notion of *programs as artifacts*: the specification, symbolic program, and physical process are to be taken as an ontological bundle that is *a program as a technical artifact*. Conceptually, we take such bundles as the output of the implementation process. This yields the following perspective on software systems as artifacts:

Requirements → *Design* → System specification → *Implementation* → Programs as Artifacts

There are, of course, many ways of carving up complex systems and conceptualizing them as technical artifacts. Requirements can take many different forms, and may themselves be expressed in different and more or less detailed ways. However, as with programs, different conceptualizations result in different artifacts.

6.5 Verification, Validation, and Malfunction

Each level of abstraction offers different possibilities for malfunction. Once the requirements are employed as a specification for the design stage, there is a possibility for malfunction. Once accepted, the design acts as a functional specification for the design of individual programs. This provides yet another possibility for malfunction. Finally, the actual physical devices may malfunction. Moreover, the conceptual natures of correctness and malfunction at each stage are, as we shall see much later, rather different.

6.6 Conclusion

The three examples in this part of the book, computers, programs, and systems, are representative of the major kinds of computational artifacts. All three examples illustrate how the abstract world of functional specifications and structural descriptions is implemented in actual physical devices. The design and implementation stages link matters and bring us, through levels of abstraction, to actual physical devices. We shall refer to these examples in almost all the discussions that follow. These cases are intended to illustrate how the notion of a technical artifact might provide a conceptual backdrop to the ontology of computer science.

Part III
SEMANTICS

The languages of computer science are formal languages defined by a formal syntax. As such they are artificial languages and, unlike natural language, there is no natural meaning attached to their expressions and terms. Consequently, their definition must go beyond their grammatical structure; they must also be given a semantic definition. This applies to all the languages of computer science whether they are specification, design or programming languages.

Our main concern in this part of the book centers on these semantic definitions and the conceptual issues that this raises. Many of these are inspired by parallel ones from the semantics of natural language and the philosophy of language. What is the role of semantic theory? Does a semantic account need to be compositional? What are the theoretical and practical roles of semantic theories? What constraints or principles must any adequate semantic theory satisfy? Should semantic definitions be given in set theory? Do we also require axiomatizations?

We shall address the majority of such questions by reference to programming and specification languages. These generate most of the semantic novelty, and suffice to illustrate the central concerns.

Chapter 7
THE LANGUAGES OF COMPUTER SCIENCE

Throughout their construction process computational artifacts are defined, specified, and described by the languages of computer science. This is one of their distinctive features. Artificial languages are employed for programming, specification, and architectural and hardware description. They are the vehicles for the expression of their functional and structural requirements. These languages operate throughout the levels of abstraction in the more global picture associated with computational systems, sometimes being employed for the expression of functional requirements, and other times for structural ones. In this chapter we provide a brief sketch of their natures, and of how they are formally defined. This will prepare the ground for a more detailed analysis devoted to their semantics.

7.1 Varieties of Languages

At the highest level of abstraction are architectural description languages. These are employed to specify the large-scale structure of computational systems and, in particular, the components and connections in a complex system. These languages include Rapide [148], Darwin [49], and Wright [9]. Garlan et al. [83] provide an analysis of the shared concepts of these languages, where the common types of objects used for representation are *components*, *connectors* and *systems*. Components generally include the primary computational elements and data stores of a system. Typical examples include clients, servers, objects, blackboards, and databases. In its simplest form a component consists of two finite sets of ports, one input set and one output set. Connectors represent interactions among components; they are the glue of architectural design. Examples include simple forms of interaction, such as pipes, procedure calls, and event broadcasts. Connectors may also represent more complex interactions such as a link between a database and an application. Systems represent configurations of components and connectors. The following is a simple example of a

© Springer-Verlag GmbH Germany, part of Springer Nature 2018
R. Turner, *Computational Artifacts*, https://doi.org/10.1007/978-3-662-55565-1_7

Wright architectural design. The components are defined in terms of their input and output ports, and the connector links the two:

CLIENT–SERVER SYSTEM
Component Server
    Port provide [provide protocol]
    Spec [Server specifcation]
Component Client
    Port request [request protocol]
    Spec [Client specifcation]
Connector C–S–connector
    Role client [client protocol]
    Role server [server protocol]
    Glue [glue protocol]
Instances
    s: Server
    c: Client
    cs: C–S–connector
Attachments
    s.provide as cs.server;
    c.request as cs.client
end SimpleExample.

At a lower level of abstraction are the languages that are explicitly termed specification languages (e.g., VDM [127], Z [258], B [6]). In their primary role, these are aimed at module and program specification. For example, in Z the central vehicle of expression is the notion of a schema. This holds two pieces of information: a declaration part that carries the type information of the identifiers of the specification, and a predicate part that is an expression in predicate logic that constrains the identifiers to satisfy the predicate. Roughly, declarations assign types to identifiers and predicates constrain them. The following would be the specification of a schema called R:

$$R = [x_1 : T_1, \ldots, x_n : T_n \mid \phi[x_1, \ldots, x_n]]$$

In VDM, specifications include preconditions, but declarations and predicates take a similar form to those of Z. We shall have more to say about such languages when we consider specification languages in a later section.

Programming languages are the core languages of the discipline [168]. They are divided into machine languages and high-level ones. The latter are classified according to their paradigm: procedural, functional, object-oriented, logical, etc. We shall shortly consider these in some detail. Of all the languages in the spectrum, they provide the most novel semantic challenge.

At the other extreme are hardware description languages (e.g., VHDL [13]). These describe the behavior of individual physical components. A digital system in VHDL consists of a design entity. Each such entity consists of a declaration part and an architecture body. The former defines the input and output signals and the latter contains the description of the entity and is composed of interconnected entities, processes and components, all operating concurrently.

Alongside these more specialized languages, there are wide-spectrum ones such as UML (Universal Modeling Language [74]). In fact, it is many different sublanguages which are divided into three categories that enable the modeling of static, behavioral, and interaction structures. Static diagrams include class and object diagrams, behavior diagrams include the use case diagrams and interaction diagrams include sequence and communication diagrams.

Outside this software development spectrum, there are query languages such as SQL. The basic structure is given by the following simple example:

```
SELECT      EmployeeID , FirstName , LastName , HireDate , City
FROM        Employees
WHERE       HireDate NOT BETWEEN '1/04/02' AND '1/11/93'
```

These are languages for structuring and querying relational databases, and are based upon predicate logic and relational algebra.

In addition, there are languages that provide the theoretical background to the discipline. Languages such as the lambda calculus and the π-calculus are employed for the mathematical investigation of computational notions. The lambda calculus [104] comes in pure and typed forms. These languages not only are aimed at the exploration of the functional paradigm, but also enable an investigation of the form and variety of the computational notion of type. Along with CSP [118], the π-calculus [166] is a language for the study of the computational notion (or notions) of a *process*, and [33] presents a language and theory of objects. These languages come with a semantic interpretation and/or axiomatization as part of their definition. And these are just the tip of the iceberg; computer science is heaving with languages.

7.2 Formal Languages

Whether they be symbolic or pictorial, the above languages are formal in nature. That is, they are defined via a precise syntax or grammar. This formality not only aids communication and avoids ambiguity, but also allows the construction of parsers, type checkers, logical frameworks, and other design tools. This is important in itself: the construction of computational systems is a complex and error-prone process, and the use of formal tools greatly aids matters at all levels.

Chapter 8
PROGRAMMING LANGUAGES

Without programming languages, machines would be idle devices much like cars without drivers or hairdressers without combs. In this chapter, we provide a guide to their definitions, and their containing paradigms. This will serve as background to one of the main objectives of this section of the book, namely to explore the semantic issues that surround programming languages.

The literature distinguishes between two kinds of programming language: *machine languages* and *high-level* ones [27]. Machine languages, as the name suggests, are formed from instructions that relate directly to the machines architecture, and are specific to a given class of machine. They have instruction sets that are executed directly by a computer's CPU where, generally, each instruction performs one operation of the machine. Typical examples include *test-and-jump*, *fetching*, *storing*, *memory reference*, and arithmetic instructions. For example, for the von Neumann machine, the *load* instruction reads a value from a memory location and *store* writes a value to one. General problem solving in these languages is difficult and unnatural because control is limited to instructions for moving data in and out of store, and *representation* is performed using numbers or Boolean values. These languages are for machines not humans. It is hard to model logical inference, the weather or a nuclear power station when all you can do is to move single digits in and out of memory. Even the programming of smart hair curlers would be tricky with these languages, to say nothing about smart football boots. Despite this, it is from these meager beginnings that high-level languages have emerged. Indeed, their very meagerness has been the inspiration for the high-level languages.

High-level languages employ more abstract concepts and control features such as procedures, abstract types, functions, polymorphism, relations, objects, classes, modules, and nondeterminism. These concepts aid problem solving, and enable a more natural representation of the problem domain. They operate at a distance from the physical machine, and do not depend upon the architecture of any specific machine. This is made possible by layers of translation and interpretation. Of course, the hard part of language design is the uncovering of such high-level concepts. And some of this uncovering is reflected in the *paradigms* or styles of programming languages [168, 79]. The main

© Springer-Verlag GmbH Germany, part of Springer Nature 2018
R. Turner, *Computational Artifacts*, https://doi.org/10.1007/978-3-662-55565-1_8

ones are the imperative, functional, logical and object-oriented paradigms. Each moves away from the physical machine in a different way, and each is characterized by the following components:

- theory of representation
- theory of computation.

The former is determined by the underlying ontology of the language, and, in particular, by its type structure, which provides the central mechanism of modeling. On the other hand, the theory of computation is largely fixed by the control features. They determine how computation [69] is realized or carried out. Our goal is to introduce these language paradigms by reflecting on these rather different ways of leaving the machine behind.

8.1 Imperative Languages

Languages that exemplify the imperative paradigm consist of commands that alter the underlying state of the machine. The term *imperative* comes from the fact that they are perceived as commands; they are instructions for the machine. Indeed, if carried out by hand, they are instructions for us. The basic instruction still involves the modification of stored values. More explicitly, the assignment statement

$$x := E$$

changes the value assigned to a location in the memory or store of the machine. Here x picks out the location and E represents the value to be inserted. Whatever value is presently assigned to x, the result of the assignment is to replace it with the result of evaluating E. In this regard, there is still a direct dependence on the von Neumann architecture. At the most basic level, the inclusion of the assignment command characterizes computation for these languages: computation changes the values stored in named locations, and more complex constructs inherit the state-based/imperative nature of assignment. Indeed, in the (sequential) imperative paradigm computation eventually unwraps into a sequence of assignment statements.

The classic language in this paradigm is Algol [14]. This was the first programming language which gave detailed attention to formal language definition: it employed Backus-Naur Form (BNF), a method of defining the syntax of formal languages. Algol has many descendants including Pascal [253] and Modula [254]. BCPL [196] and C [133] are also imperative. But, to illustrate matters, we shall employ a language that in some form is at the core of most of these languages. The WHILE programming language [48] is a small imperative language whose syntactic definition, the grammar of the language, is given as follows:

$P::=$skip$|x:=E|P;P|$if E then P else $P|$while E then $P|$

$E::= x \ |0|1|E+E|ExE|$true$|$false$|E=E|E<E|\neg B|B \wedge B|$

Simple programs are constructed via a skip operation that does nothing to the state, and the assignment statement $(x:=E)$ itself. Complex programs are generated from these by sequential composition (;), conditionals (if then else), and iteration (while do). Atomic expressions contain variables x, 0, and 1, and Boolean values (true and false). Expressions are closed under addition, multiplication, less than, negation ($\neg$) and conjunction ($\wedge$).

The following is a program that increases the value of x by 1, and changes the value of y to that of the present value of x times the old value of y itself:

```
x := x + 1; y := x * y
```

The above syntactic definition of the actual language is a recursive definition. For example, in the *while* command while B do P we may use any expression or program that has already been constructed to replace B and P. The following illustrates this. The first two lines are atomic statements, but the next imports the above program into the *while* construct.

```
x := 0;
y := 1;
while x < n do (x := x + 1; y := x * y)
```

The theory or model of computation for this paradigm may be summarized as follows:

- Programs are operations on states, and computation is state transformation.

We have not strayed too far from the machine: the underlying model of computation is not unlike that of machine code itself. However, problem solving is aided by the use of constructs such as iteration and conditionals. These enable the programmer to express the problem-solving strategy at a conceptual distance from the operations of the machine. In particular, iteration, in the form of the *while* command, represents an early instance of abstraction, where the exact sequence of machine instructions is hidden from view; that sequence appears only when the complex commands of the language are executed.

The imperative paradigm is often called *procedural* because these languages also include procedures that abstract the code of a program into a named structure that may be called with different arguments. For example, the following introduces a named procedure that takes three arguments. When it is applied or called, it is supplied with actual arguments (a, b, c), and these replace x, y, z in the body P:

$$p(x, y, z) = P.$$

In addition, to enhance their representational mechanisms, high-level languages have a richer collection of *data types*. The type structure of the language provides its fundamental means of modeling. In the early days, the data types of these languages (e.g., stacks and arrays) stayed close to the structure of the underlying machine. However, there is no conceptual reason why a more mathematically inspired type structure might not be attached to them. For example, our simple imperative language might have the following data types:

$$T ::= N|Bool|T{\otimes}T|Set(T)$$

Here the basic types are either numbers or Booleans. More complex types are built using type constructors for Cartesian products and finite sets. These come with additional syntactic operators such as those for ordered pairs and set-theoretic union and intersection. Such types provide a more expressive means of representation. For example, products are the types of ordered pairs and, together with finite sets, facilitate the representation of finite relations. This enables the employment of tables that, for some given types A and B, are members of the type $Set(A \otimes B)$. In general, the richer the class of type constructors, the closer one can get to the structures and systems that are to be modeled.

The languages of this paradigm represent the first attempts to design high-level languages. Moreover, to many this imperative style of programming is still a natural way of programming. Indeed, much coding is still done in the imperative style, where the implicit model of computation, the way in which computation proceeds, is essentially that of the underlying physical machine. Presumably, this is taken to be natural because programming is seen as the giving of a sequence of instructions. This paradigm owes much to Turing [229].

However, there are many perceived drawbacks. The central one concerns the correctness of programs written in the language. Reasoning about imperative programs involves reasoning about the underlying state; one has to reason about the way the state changes. This may have little to do with the problem being solved, where generally this does not make reference to any such underlying state. Moreover, this appears to be quite different from the kind of reasoning to be found in mathematics, the ideal for formal reasoning. Indeed, the next paradigm was introduced to address this issue.

8.2 Functional Languages

In the *functional* paradigm computations are expressed as the evaluation of mathematical functions. Languages such as (pure) Lisp [94, 217], Miranda [227], Haskell, [226] and SML [105] are examples. In a pure functional language there is no assignment, and no side effects of computation; programs are evaluated as mathematical functions.

If the WHILE language is at the heart of the imperative paradigm, the lambda calculus, together with its combinatorial variants, is at the heart of the functional one. The lambda calculus [104] is a pure language of functions that forms the basis of the Church-Kleene account of formal computability. The syntax is minimal:

$$t ::= x | \lambda x.t | tt$$

Lambda terms are either variables (x), abstraction terms $(\lambda x.t)$, or the application of one term to another (tt). In programming terms, abstraction is (unnamed) function definition, and application is function application or call.

The underlying model of computation is *reduction*, where functions in the form of lambda abstractions are applied to their arguments. The reduction is instigated by substituting the argument for the variable of abstraction.

$(\lambda x.t)s$ reduces to $t[s/x]$

i.e., t with s substituted for x. The lambda calculus, as a formal system, has rules governing functional abstraction and application that reflect this model. Reasoning about such computations involves no notion of state; it involves only these reduction rules. Peter Landin [141] used the lambda calculus as the basis for a form of (operational) semantics for all programming languages. Unfortunately, as a programming language it is as impractical as machine code. But it does illustrate the underlying principles that govern this paradigm:

• Programs are (mathematical) functions and computation is the evaluation of functions.

This language and its combinatorial variants form the basis of functional languages. For example, the following is a simple Haskell program involving function definitions and applications. The first line defines the function *Squarish* and the rest applies it:

```
Squarish (a, b) = a * a - b * b
len vec = sqrt (Squarish vec)
main = print $ len (19, 15)
```

Of course, realistic functional languages have a much enriched type and operational structure. Indeed, in the functional paradigm the development of typed languages has been prominent. To illustrate matters we consider the simple typed lambda calculus [18]. The types are given as follows: types are either basic (B) or the type of functions from one type to a second:

$$T ::= B|T{\Rightarrow}T$$

The language is given by the following simple rules, where d is a set of type assignments to variables:

$$\frac{d; x : T \vdash t : S}{d \vdash \lambda x.t : T \Rightarrow S} \qquad \frac{d \vdash t : T{\Rightarrow}S \quad d \vdash s : T}{d \vdash ts : S}$$

The first introduces the abstractions, but now they are typed. The second rule introduces applications, where the type of the function's domain must match that of the argument. These rules also implicitly serve as rules of term formation; they provide the grammar of the typed calculus. BNF has been usurped by type introduction and elimination rules.

Generally, functional languages have contributed more to type theory than the other paradigms have. In particular, they pioneered the development of the various notions of polymorphism [195] and abstract types. And, generally, in a manner similar to dimensional analysis in physics, languages employ types to partially ensure the correctness of programs. On the operational side such languages contain explicit mechanisms such as recursion that enable an elegant form of inductive reasoning about their properties.

While functional languages are said to aid correctness, the advertised downside concerns the inefficiency that surrounds their implementation. One drawback concerns the way that storage is maintained. In particular, it is no longer controlled by the user of the language; the user has handed over control to the implementation. If a storage location is used to hold a value, it is not by itself available for reuse. Consequently, a separate part of the implementation must determine whether the location is no longer being used. However, this inefficiency is probably not the only reason these languages have found little application in the commercial and industrial world. Indeed, the power of modern hardware has to a large extent addressed this concern. Unfortunately, functional languages are often taken to be too mathematical for real applications. Paradoxically, the fact that this is their main advantage also appears to be their main drawback. Whether this is a consequence of the poor mathematical education of computer scientists is open to debate. Fortunately, these languages are undergoing a renaissance.

This paradigm owes much to Church [36].

8.3 Logical Languages

The *logic* paradigm represents yet another way of moving away from the imperative style. Here the basic concept is the logical notion of assertion or proposition [80, 172]. One way of understanding this approach is in terms of specification. Very roughly, a specification states what a program must do, not how one should do it. The idea behind logic programming, at least a simple reconstruction of the central idea, is to move specification and programming closer together by employing a programming language that is close to being a specification language. The general idea is to be able to state the problem rather than solve it – a task left to the implementation of the language. In this paradigm programs are assertions or goals.

More specifically, the underlying language of logic programming is a fragment of predicate logic called *Horn clause logic*, the syntax of which is given as follows:

```
H ::=  p:−A
A ::=  p | p;A
p ::=  R(t1 ,... , tn )
```

A Horn clause (H) is an assumption list (A), which is a list of atomic propositions, followed by an atomic proposition (p), the conclusion. The assumption list describes what we know, and the conclusion is the goal. In logical terms, we may think of a Horn clause as an inference rule of the following form.

$$\frac{p_1, \dots, p_n}{p}$$

Another way of looking at matters is to regard the assumption list as the known database, and the conclusion as a query to be evaluated against it. The following is a simple set of clauses:

```
drinksredwine (Snod).
playsgames (Chris).
playsgames (Snod):− drinksredwine (Snod).
playscards (Chris):− playsgames (Chris).
playscards (Snod):− playsgames (Snod).
```

If a knowledge base contains the clause *head :- body*, and we know that the body follows from the information in the knowledge base, then we may infer the head. For example, given the above set of assumptions, we can deduce "playscards(Snod)". Computation in this paradigm is a truncated form of theorem proving called unification [80]. So the fundamental principles that determine this paradigm are the following:

- Programs are collections of logical assertions, and computation is inference.

Prolog [172] is the main language to emerge from this paradigm. Strangely, this paradigm has generated fewer languages than have the other paradigms. Nor has it contributed much to type theory. However, it does attempt to raise programming to the level of specification.

This paradigm owes much to Frege and Russell, the inventors of modern logic.

8.4 Object Orientation

Finally, object-oriented programming (OOP) views the world as a collection of *objects* and *classes*. Java [200] and Eiffel [66] are two members of the genre.

The central idea is easy enough to grasp. Consider cars. How may we describe the notion of a car? Presumably, it has properties such as the make, model, engine, etc. These properties are taken to determine the notion of a car. In addition, we can do things with cars, and they can even do things to themselves. We can turn the car on or off, we can accelerate or brake, etc. In OOP, these actions are called *methods*. They allow the objects to do things, and enable the programmer to manipulate the properties of an object. We can lump all this information together in the description of a class of objects called *car*:

Car
make: String , model: String , engine :Num
Find Speed , Start , Stop .

One of the main tenets of this style is *encapsulation*: everything an object requires should be part of the object i.e., its attributes and the methods that maybe employed to modify it should be part of the class definition. Classes inherit properties and methods from their *superclasses*. Since objects have methods, computation may be induced by objects sending messages to each other. So, an object called *driver* may send an operational message *start* to an object called *car*.

The fundamental principles that determine this paradigm are the following:

- Programs are objects, and computation is message passing.

The latter is generated when one object sends a message to another to invoke one of its methods. This is the paradigm of Java and Eiffel, and at the time of writing this book, it is probably the dominant one.

The main claim of this paradigm is *naturalness*, and the ease of representation that follows in its wake. *Naturalness* relates to the fact that it facilitates the representation of real-world objects which can be conceptualized via their attributes and operational methods. In other words, there is an underlying claim to the effect that a *natural* representation of everyday notions is an object-oriented one. Generally, aside from appeal to common sense, and experience of use, not much empirical evidence or conceptual argumentation is offered for such a claim. But much the same might be said for the claims of the other paradigms.

This paradigm owes much to Aristotle, the inventor of pre-Fregean logic.

8.5 Concurrency and Nondeterminism

At the core of these paradigms lies their model of computation: *state-based, functional, unification,* and *message passing*. However, there is also another axis of comparison, namely whether computation is *sequential* or *concurrent*. In sequential computation, individual components of the language are executed in sequence, one after another. In a concurrent program, several streams of operations may operate at the same time. Each stream executes sequentially, with the additional feature that streams can pass messages and interfere with one another. Each such sequence of instructions is called a thread. Controlling the interactions between different threads and shared resources provides some challenges. *Race conditions, deadlocks,* and *resource starvation* are common issues.

The theoretical languages used to model and explore concurrency include CSP [118] and the π-calculus [166]. Most modern languages have variants that allow the programmer to express concurrent computations, and all the major paradigms contain languages that support some form of concurrency. Concurrent Haskell and Concurrent ML are functional whereas Shapiro's Concurrent Prolog [203] is logical. Concurrent Pascal is imperative while concurrent Java is object-oriented.

A further dimension concerns the *deterministic/nondeterministic* distinction. A nondeterministic program may, for the same input, exhibit different behaviors on different executions. For example, a concurrent program can perform differently on different runs owing to a race condition. Conceptually nondeterminism may be illustrated by the use of so-called guarded commands – a generalization of the conditional: the guards choose which arm to execute. If more than one guard is true, one statement is "nondeterministically" chosen to be executed. For example, in the following G_i are guards, and the S_i fire when the corresponding guard is fired:

if $G_0 \to S_0 | G_1 \to S_1 | \ldots | G \to S_n$ fi

Many contemporary languages include some such features.

8.6 Theories of Representation and Computation

Paradigms provide ways of moving away from the physical machine. They provide languages that enable problems to be solved using a means of representation that is closer to the ontology of the problem domain, and hopefully closer to the way we solve problems. At least, this is the aim and claim of paradigm-based language design. However, they achieve this in different ways.

Functional languages insist that everything should be modeled as functions, with functional application as computation. This is inspired by the approach to representation that employs sets and functions. The main concern is to enable the construction of correct programs, and they achieve this by programming with a core mathematical notion, namely that of a function. In principle, the state is invisible to the functional programmer. Generally, these languages supply a rich language of types and type constructors that enable a variety of representational devices.

Logical languages demand a propositional representation, with deduction as computation. The aim is to bring programming closer to specification. This paradigm has contributed less to the representational side of languages than the functional one has: type theory has not been a major design issue.

The object-oriented model insists that the whole world can be modeled as classes and objects; it insists that the natural way of representing real-world notions is in terms of attributes and methods. Message passing and interaction are its means of computation.

These paradigms represent radically different approaches to representation and computation. Each has its own ontology: functions, propositions, and objects; and its own notion of computation: function application, deduction, and interaction. And many languages are mixed. No doubt the future will see others.

Paradigms are motivated by an implicit *theory* about the best way to solve problems. But is there any real justification for these different theories? Is it even clear what such claims about problem solving amount to, and how to test them? Are there advantages to be found in different applications? What really distinguishes functional reasoning from state-based reasoning? Why is object orientation *natural*? While we shall eventually address some of these questions, in this part of the book, we shall be more concerned with the logical nature of these languages, and in particular their semantic definitions.

Chapter 9
SEMANTIC THEORIES

Aside from its syntax, a programming language must also be given a semantic definition and an implementation. Without a semantic definition, we could not compute by hand; without an implementation, there would be no mechanical computation. These three ingredients, syntax, semantics, and implementation, are all necessary, and work together to define a programming language.

Users of programming languages require some account of the intended meanings of the constructs of the language, where, presumably, the intentions are those of the language designer. Such accounts come in all shapes and sizes. In practice, natural language accounts are common. However, programming languages employ vernacular terms in more exacting ways, and introduce new technical notions that need to be precisely defined. This has led computer scientists to create a whole range of more formal options. The literature here is vast and varied, but the major players in the semantics stakes are split into two main camps: denotational [92, 204, 218, 213, 224, 251] and operational approaches [68, 189], with game-theoretic versions of the former playing a significant role [2, 4]. There are also axiomatic systems that link programming constructs with logical descriptions of their intended impact [51, 114]. We shall eventually discuss all these approaches, but here we introduce the desideratum for a semantic account: what must a semantic account do?

9.1 The Roles of Semantics

In order to design a program, or to establish or explain why it satisfies its specification, a programmer must reason about the properties of the language and its constructs. For this, she needs to know the semantic impact of each construct in the program.

To illustrate matters, consider the following program. The name gives the game away: it is intended to compute the factorial function.

© Springer-Verlag GmbH Germany, part of Springer Nature 2018
R. Turner, *Computational Artifacts*, https://doi.org/10.1007/978-3-662-55565-1_9

```
Function Factorial(n : Integer) : LongInt;
Var
     Result : LongInt;
     i : Integer;
Begin
     Result := n;
     If (n <= 1) then
     Result := 1
     Else
        For i := n-1 DownTo 1 do
        Result := Result * i;
Factorial := Result;
End;
```

In order to construct or even understand this program the programmer must know the impact of assignment statements, sequencing, conditional expressions and iteration. And she must somehow use this knowledge to reason about their combined impact. In actual fact, there are several places in the construction and the verification stages of programming where semantic knowledge is necessary.

Suppose that a programmer is faced with the task of designing a program that computes the factorial function. Suppose further that there is no independent semantic account of the language; the user knows only the syntax of the language. We may assume that she can use the implementation to execute her programs, but has no knowledge of its internal workings. How could she design such a program? She might try aiming well-formed but arbitrary strings of symbols at the machine, running them and observing their behavior. Generally, such a hit-and-miss approach would render programming impossible. Of course, she might be able to guess some of the semantic content of the language, at least roughly. For instance, she might use her knowledge of the English words that occur (e.g. *for*, *if*, *then*) to gain some foothold. While this might help, it would still be a very hit-and-miss affair, and subject to ambiguity and misinterpretation. The words that occur in programming languages do not just inherit their vernacular senses; they are technical terms whose meanings need to be separately and precisely spelled out. Of course, she could use the implementation to extract some kind of meaning, but this would be no easy task, and she would never be sure that it correctly captured the intention of the language designer. And, even if she succeeded, she would now be using a reverse-engineered semantic definition to construct programs.

Semantic knowledge is required not just to design, construct, or understand programs. It is also necessary in order to argue that a program works, satisfies its specification. To do so one must link the overall impact of the program with its functional specification. The programmer must also understand the specification language, its semantics, and the semantic links between the languages.

In practice, this is often done in a very informal way, but this does not mean that semantic knowledge is not employed in the process. The correctness or verification process involves reasoning, and for this semantic knowledge is required.

Allied with this is the challenge of explaining why a program does what it is supposed to do. Asked to say why a program works, not if it does, but why it does, the programmer needs to be able to explain matters. And such explanations must refer to the meanings of the individual components. In the case of a high-level language, for practical reasons alone, these explanations could not employ the details of the underlying implementation.

Designing, establishing, and explaining are rational activities that must appeal to the semantic knowledge of the language. This does not imply that experimentation is not part of the program construction process, nor that it may not be part of some verification process. It only insists that reasoning about the meaning of the constructs must form an essential part of the whole process of program construction, correctness, and explanation. This concerns the programmers use of the language. But semantics must also guide the implementation of the language; it must reflect the intentions of the designer about the constructs of the language. Consequently, it must provide the functional specification of any implementation: an implementer must implement a language in harmony with its semantic interpretation.

For both programmers and implementers, semantic descriptions provide a *notion of correctness*. This semantic requirement has its origins in the philosophy of language.

9.2 Normativity

Although the exact notion of normativity for semantic accounts in the philosophy of language is complex [88], there appears to be a minimum requirement on which most are agreed:

> The fact that the expression means something implies, that there is a whole set of normative truths about my behavior with that expression: namely, that my use of it is correct in application to certain objects and not in application to others. The normativity of meaning turns out to be, in other words, simply a new name for the familiar fact that, regardless of whether one thinks of meaning in truth-theoretic or assertion-theoretic terms, meaningful expressions possess conditions of correct use. Kripke's insight was to realize that this observation may be converted into a condition of adequacy on theories of the determination of meaning: any proposed candidate for the property in virtue of which an expression has meaning, must be such as to ground the "normativity" of meaning – it ought to be possible to read off from any alleged meaning constituting property of a word, what is the correct use of that word [24].

To say that meaning is essentially normative is to say that certain norms are valid, or in force. In terms of programming languages, the "normativity of meaning" has it that any semantic account must provide a criterion of correctness. However it is expressed or conveyed, the semantics of a language must guide all categories of users: it must inform the user when she has correctly used a construct of the language and when she has not, and it must inform the implementer when she has correctly implemented a language and when she has not. In particular, it must facilitate a specification of compiler correctness, where a compiler has to translate between two languages. Any such translation must preserve the semantic interpretation of the source and target languages.

This is the correctness or normative constraint on semantic definitions. For this to be so, there must be agreement of judgments about the content of any such semantic account. This is particularly true for artificial languages. The semantic definition of the language must be such that users of the language (e.g., programmers and implementers) are able to grasp and employ it, and there must be agreement in the community of users about its content. So, it must be expressed in a language that is shared.

9.3 Compositionality

A further demand on programming language semantics also has its origins in the philosophy of language: a semantic account must be compositional. For a natural language, this is taken to be a constraint upon our grasp of the language: it is taken as a constraint on the relationship between linguistic structure and meaning that underpins how understanding and learning a language is possible. In its most straightforward form, the principle of compositionality may be stated as follows:

> For every complex expression E in L, the meaning of E in L is determined by the structure of E in L and the meanings of the constituents of E in L [221].

For programming languages, the compositional view insists that the meaning of a complex program is fully determined by its linguistic structure, and the meanings of its constituents. In other words, the structure, together with the meaning of the building blocks, fixes matters.

Frege [76] claimed that the possibility of our understanding sentences we have never heard before depends on our ability to construct the sense of a sentence from the senses of its constituent parts. This is the argument from productivity and comprehension. In particular, competent speakers are able to understand a sentence or expression E they never encountered before. To do so, they must possess knowledge that, by itself, enables them to grasp the meaning of E. Presumably, this can only be knowledge of the structure of E together with knowledge of the individual meanings of its constituents. This is taken to be the best explanation of our creative linguistic ability; it is an inference to the best explanation [221]. At this level of generality, it is hard to dispute these observations. It is a general argument for compositional semantics that would seem to be applicable to all languages, natural and artificial.

However, there are more specific ones for programming languages that, if anything, strengthen the demand for a compositional semantics. One account of compositionality that is especially suitable to the present setting identifies a compositional interpretation of programming languages with a homomorphism from the algebra of program phrases to the algebra of the mathematical structure that supplies its semantic interpretation. Tennent [223] suggests some reasons for using such denotational definitions:

> In a denotational definition, each phrase of a language is given a meaning that describes its contribution to the meaning of a complete program that contains it. Furthermore, the meaning of each phrase is formulated as a function of the denotations of its immediate sub phrases. As a result, whenever two phrases have the same denotation, one can be replaced by the other without changing the meaning of the program. Therefore a denotational semantics supports the substitution of semantically equivalent phrases.

A compositional semantics allows the substitution of semantically equivalent parts. Provided that we know that a semantically well-defined unit is denotationally equivalent to another, we can substitute it *salva veritate*. This facilitates modular program design. We shall have more to say about this when we discuss programming itself. But the main thrust of the argument seems clear: a compositional semantics enables the substitution of equals for equals, and this supports modular design. This is not an argument about comprehension but about the use of the language: languages that have a compositional semantics are easier to use; they aid the process of programming:

A second aspect relates directly to establishing properties of the language.

> Since a denotational definition parallels the syntactic structure of its BNF specification, properties of constructs in the language can be verified by structural induction, the version of mathematical induction that follows the syntactic structure of phrases in the language [223].

Structural induction is the main mechanism for systematically reasoning about the language as a whole: we show that the property holds for the atomic statements of the language, and show that it is preserved by a complex one. And such inductions are justified by compositional semantics. In other words, proving properties of programs is facilitated by compositional semantic accounts.

The last of the three arguments is aimed at the design process for languages. It is argued that, since it enables the evaluation of the impact of different constructs in isolation from the rest of the language, compositional semantics supports the design of languages.

> Compositional semantics lends a certain elegance to denotational definitions, since the semantic equations are structured by the syntax of the language. Moreover, this structure allows the individual language constructs to be analyzed and evaluated in relative isolation from other features in the language [223].

The impact of the inclusion of individual constructs can be more directly evaluated when the semantic definition of the language is compositional.

In summary, it is argued that semantic definitions that are compositional have merit in that they support comprehension, productiveness, explanation, and correctness. In addition, they aid both language and program design.

9.4 Rigor

Earlier we alluded to the use of natural language in semantic accounts. There are various ways in which this might occur. One possible use concerns the employment of natural language terms to stand proxy for the technical ones of the artificial language. For example, terms such as *if, then, until, name, value, program, object, type, expression, catch, throw, command, parallel, location* are used in or about programming languages. If these words inherit their natural language meanings, then the programmer could employ their vernacular meaning. Since programmers have a working understanding of these terms, this may guide and inform their use in programming contexts. However, when used in programming contexts such words become technical terms, and their meaning cannot simply be taken to be their vernacular one. We are warned of this by one of the founders of semantic theory:

> Any discussion on the foundations of computing runs into severe problems right at the start. The difficulty is that although we all use words such as 'name', 'value', 'program', 'expression' or 'command' which we think we understand, it often turns out on closer investigation that in point of fact we all mean different things by these words, so that communication is at best precarious. These misunderstandings arise in at least two ways. The first is straightforwardly incorrect or muddled thinking. An investigation of the meanings of these basic terms is undoubtedly an exercise in mathematical logic and neither to the taste nor within the field of competence of many people who work on programming languages. As a result the practice and development of programming languages has outrun our ability to fit them into a secure mathematical framework so that they have to be described in ad hoc ways. Because these start from various points they often use conflicting and sometimes also inconsistent interpretations of the same basic terms [219].

Words such as *name, value, catch,* and *throw* are natural language terms, and while their vernacular meanings may act as a guide to the user, as technical terms they have a semantic life of their own. Vernacular interpretation often leads to different interpretations by different users, and may result in inconsistency. For example, the word *evaluate* has a reasonably precise natural language meaning, but we cannot just import the vernacular word *evaluate* to unpack the content of the following:

evaluate P in state s.

where P is some program in a given language. The natural language term does not tell us how, given the results of the evaluation of their components, to evaluate complex expressions. In itself it does not provide a semantic account of any kind, let alone a compositional one. Programming requires that the technical terms of the language be defined; they cannot just inherit their natural language meaning. While many learn programming languages by the use of metaphors [123] such as *catch* and *throw*, these are psychological props that are part of the learning process [43]. But it would be a useless mistake to base any kind of semantic theory on these notions. This is an argument against the simple-minded notion that semantics should employ technical terms as if they were natural language terms. However, this is not to say that natural language cannot be employed in the semantic enterprise. We shall see how shortly.

9.5 Conclusion

This chapter provides some account of the conceptual questions that arise with the semantic definition of programming languages. There are others that involve more technical aspects of formal semantics, but these bring in issues that are more naturally addressed within theoretical computer science or the philosophy of mathematics. There are also related issues that pertain to the design of such languages, but we need to cover more ground before we can address these.

Chapter 10
FORMAL SEMANTICS

In this chapter we look more carefully at the various forms of semantic definition [26, 68, 92, 97, 114, 141, 167, 173, 189, 204, 213, 218, 219, 224, 251, 246]. These range from natural language accounts through to mathematical ones of various kinds and flavors. We shall attempt to evaluate the various approaches to semantics against the criteria set out in the previous chapter. In this regard, we shall explore the different roles of operational and denotational approaches.

10.1 Vernacular Semantics

To illustrate matters, we employ a semantic description of the WHILE language. Any semantic definition for this language must begin with some account of some abstract machine with an underlying notion of state. For simplicity, we adopt the following system:

1. A state is a mathematical function from a set of *locations* L to a set of numerical *values* V.
2. There is an operation (*Update*) that, given a state s, a location l, and a value v, changes the state to one where the location l is assigned the value v.

We might express matters in other ways, using different mathematical notions, but this simple machine will be sufficient to illustrate the main ideas. In what follows, we define the evaluation of programs relative to the states of this machine. Formally, we define the following relation:

Evaluating the program P in state s terminates in the state s'.

The following provides a form of operational semantics. It is usually called *big step semantics*, and is given in terms of rules that provide the evaluation of a complex program in terms of the evaluation of its parts:

1. To evaluate **skip** in a state s, return s.

© Springer-Verlag GmbH Germany, part of Springer Nature 2018
R. Turner, *Computational Artifacts*, https://doi.org/10.1007/978-3-662-55565-1_10

2. To evaluate $x := E$ in a state s, evaluate E in s and place the resulting value in location x, overriding any existing values in this location, i.e., $Update(s, l, v)$.

3. To evaluate $P; Q$ in state s, first evaluate P in s. If this returns a state s', evaluate Q in s'.

4. If the evaluation of B in s returns true and the evaluation of P in s returns s', then the evaluation of if B then P else Q in s evaluates to s'. If, on the other hand, the evaluation of B in s returns false and the evaluation of Q in s returns s' then the evaluation of if B then P else Q in s evaluates to s'.

5. If the evaluation of B in s returns true, the evaluation of P in s returns s', and the evaluation of while B do P in s' yields s'', then the evaluation of while B do P in s returns s''. If the evaluation B in s returns false, then the evaluation of while B do P in s returns s.

We have defined *evaluating the program P in state s terminates in the state s'* by a definition that unwinds through the structure of the language. The word *evaluate* is a technical word that has its meaning given compositionally, following the syntactic structure of the language. It is parallel to the Tarski definition of truth for logical languages. The great advantage of natural language as a semantic vehicle is that it is shared and understood. Where natural language terms are used, for example, *and, then*, etc., their meaning is taken to be sufficiently precise to fix the precise meaning of the technical term *evaluate*.

However, even though the definition is expressed in vernacular language, it is a mathematical definition. And this is so in the same way in which ordinary mathematics is carried out in a mixture of formal and natural language. Moreover, the above semantic clauses provide us with a semantic account that does not depend upon physical properties. On the face of it, the above semantic definition satisfies all our normative and compositional demands. Moreover, it is carried out in a language that supports agreement of judgment between users. For this reason alone, such accounts accompany more formal ones, usually as a commentary.

Nevertheless, there are aspects that are buried, or at least not made explicit, in the informal account given above. First, observe that the definition is *recursive* in the sense that the evaluation of complex programs gives rise to further reference to the whole evaluation process. This is part and parcel of its compositional nature. But how is this recursion justified? This is a technical question that cannot be adequately addressed without further mathematical investigation and support.

10.2 Operational Semantics

With reference to the language of mathematics and the reasoning that mathematicians engage in, Frege wrote:

> The logical imperfections of language stood in the way of such investigations. I tried to overcome these obstacles
> with my concept-script. In this way I was led from mathematics to logic [76].

Hilbert [111] and Frege [76] provide the origin of modern axiomatic systems with their precise rules
of inference. Indeed, it is this tradition that influenced much of the early work in formal semantics.
In this section, we shall provide a more formal version of the above vernacular semantics. This will
enable us to address the kind of mathematical concerns raised at the end of the previous section.

To enable a more mathematically explicit formalization, we adopt some notation. We shall write

$$< P, s >\Downarrow s'$$

to express the judgment that evaluating the program P in state s terminates and returns the state
s'. With this to hand, we may express the semantic clauses in rule-based form. Here we employ
premise/conclusion rules to reflect the conditional nature of the informal semantic clauses:

$$\frac{Premise\dots Premise}{Conclusion}.$$

1. For the skip command, the semantic rule unpacks as follows. This is the atomic step in the
   recursion; the skip step does nothing:

$$< skip, s >\Downarrow s$$

There are no premises to the rule, since it is an atomic instruction, and there is no dependent
evaluation.

2. The assignment statement requires a prior evaluation of the expression on the right-hand side.
   If the expression evaluates to some value v, then, in the conclusion, we replace the value of the
   variable with the value of the expression in the current state:

$$\frac{< E, s >\Downarrow v}{< x := E, s >\Downarrow Update[s, x, v]}.$$

Unlike the skip command, which is written as an axiom, this requires a rule where the premise
corresponds to the antecedent of the conditional in the informal rule. Here we assume that
evaluating expressions does not change the state.

3. The sequencing of programs is unpacked exactly as in the vernacular account, as relational
   composition:

$$\frac{< P, s >\Downarrow s' \qquad < Q, s' >\Downarrow s''}{< P; Q, s >\Downarrow s''}$$

4. For conditionals, our natural language account is transformed as follows. Notice that the two assumptions correspond to the antecedent of the conditional in the natural language version:

$$\frac{< B, s >\Downarrow \text{true} \qquad < P, s >\Downarrow s'}{< \text{if } B \text{ then } P \text{ else } Q, s >\Downarrow s'}$$

$$\frac{< B, s >\Downarrow \text{false} \qquad < Q, s >\Downarrow s'}{< \text{if } B \text{ then } P \text{ else } Q, s >\Downarrow s'}$$

Again, the transition from the informal to formal rule is straightforward.

5. Finally, we consider the most complex case involving the iteration. In the rule-based account, the semantic rules split according to whether the Boolean is true or false. The formal account follows suite:

$$\frac{< B, s >\Downarrow \text{false}}{< \text{while } B \text{ do } P, s >\Downarrow s}$$

$$\frac{< B, s >\Downarrow \text{true} \qquad < P, s >\Downarrow s' \qquad < \text{while } B \text{ do } P, s' >\Downarrow s''}{< \text{while } B \text{ do } P, s >\Downarrow s''}$$

It should be fairly clear that our formal description is nothing more than a reformulation of the natural language one, where

$$< P, s >\Downarrow s'$$

expresses the statement that *evaluating the program P in state s terminates in the state s'*, and the informal statements are recast as rules with the antecedent of the informal rule functioning as the premise, and the consequent as the conclusion. All we have done is make the inferential structure of the rules explicit. All this is not that far removed from the informal account, but it is more wholesome and less prone to misunderstanding: the informality of the natural language account has been made more precise, explicit, and systematic. However, usually both formal and informal accounts are given side by side.

The rules for the evaluation of expressions take a similar form, where the basic judgment is

$$< E, s >\Downarrow v$$

which expresses the judgment that evaluating the expression E in state s returns the value v.

There is also a small-step operational semantics with a basic relation

$$< P, s >\Downarrow < P', s' >$$

which encodes the idea that in a single computation step the program P is transformed into P' and the state s is transformed into s'. These two accounts are equivalent in the sense that, that from a given state, the small-step semantics reaches a final state exactly when the big-step semantics yields a result.

We can now put this formalism to work. In particular, we can more easily show that for each program and each input state there is at most one final state:

$$\forall P.\forall s.\forall s'.\forall s''.(< P, s > \Downarrow s' \wedge < P, s > \Downarrow s'') \rightarrow s' = s''.$$

This is a theorem of the system, which we may prove by induction on the depth of the proof trees. For the induction step, we assume that the above holds for the premises of the rule, and show that it holds for the conclusion. This is an induction on the derivation structure, not on the structure of the language. A structural induction would not work, since the premise of the last *while* rule makes reference to the *while* construct itself, and so there is no reduction in the inductive complexity. Instead, we need to carry out induction on the derivation trees. For this, the tree-like structure of the derivations must be made explicit, as it is in the formal account.

Formal semantics must act as a normative guide for users. It can only do this if the semantic account is understood and agreed by users of the language. The fact that most semantic accounts are accompanied by their natural language variants facilitates this. Some will only appeal to the formal account where there is a lack of clarity. At the very least, formal semantics is an essential tool for exploring and establishing the properties of languages, and for guaranteeing that the informal accounts are mathematically coherent.

In the lambda calculus, the above theorem would be called a *Church-Rosser* property, and it is a required consistency condition for sequential languages. It guarantees that, with the same starting conditions, the evaluation process, if it yields a result, always yields the same one. This means that we may associate with each construct a partial function from states to states as follows:

$$[P]s = \left\{ \begin{array}{l} s' \text{ if } < P, s > \Downarrow s' \\ \text{undefined otherwise} \end{array} \right\}$$

Observe that we can make the compositional nature of the semantics more explicit by rewriting the rules using this functional notation.

$$\frac{[P]s = s' \qquad [Q]s' = s''}{[P; Q]s = s''}$$

i.e., $[P; Q] = [P] \circ [Q]$. So that, sequencing is compositionally interpreted as functional composition. The same is true for all the constructs, and this leads naturally to the denotational approach.

10.3 Denotational Semantics

The denotational semantics for programs is given by defining, for a program P, a partial function that, given an input state s, returns the state that results from executing P in the state s. Of course, there is the possibility that P fails to terminate when it is executed in s, in which case it will not return any state: in other words, the interpretation must be a partial function from states to states. For the constructs we have discussed, the semantics takes the following form, where $\simeq$ is a partial equality relation that allows for the fact that programs may not terminate:

1. The skip instruction is self-evident: the state is unaffected:

$$\|skip\| \, s \simeq s.$$

2. Assignment updates the machine as previously indicated:

$$\|x := E\| \, s \simeq Update(x, \|E\| \, s, s).$$

3. Sequencing is functional composition:

$$\|P; Q\| \, s \simeq \|Q\| \, (\|P\|)s.$$

4. Conditionals also unpack as one would expect from the operational account:

$$\|\text{if } B \text{ then } P \text{ else } Q\| \, s \simeq \left\{ \begin{array}{l} \|P\| \, s \text{ if } \|B\| \, s = \text{true} \\ \|Q\| \, s \text{ if } \|B\| \, s = \text{false} \end{array} \right\}.$$

5. Finally, the functional semantics for the *while* loop is given as follows.

$$\|\text{while } B \text{ do } P\| \, s \simeq \left\{ \begin{array}{l} s \quad \text{if} \quad \|B\| \, s = \text{false} \\ \|\text{while } B \text{ do } P\| \, (\|P\| \, s) \quad \text{if} \quad \|B\| \, s = \text{true} \end{array} \right\}.$$

This yields a version of denotational semantics where each program is associated with a partial state-to-state (set-theoretic) function.

Notice how the last clause of the definition, the *while* clause, requires some justification since there is a hidden *fixed point* construction. Indeed, while this issue is made explicit in the denotational version, it is already implicit in the operational one. On the face of it, the definition is circular: the *while* construct occurs on both sides of the semantic clause. It requires some mathematical unpacking: we need to know that such definitions coherent. For this we need some underlying theory, such as partial domain theory [189], that supports the construction of *fixed points* for continuous functions over the topology of the partial domain theory. This requires mathematical work. But the technical details are not our main concern. What is, is the fact that the issue is brought to the surface, and can only be properly addressed with the appropriate mathematical formulation. This is a foundational argument for the necessity of a formal semantic account that is cast within a mathematical theory that supports the required mathematical properties. So, the argument for a more formal articulation of the semantics is mathematical in nature. More explicitly, there are mathematical requirements on the informal version that need to be uncovered and resolved, and these are best addressed with a degree of formalization.

10.4 Definitional Priority

We can now address the question of the relationship between these various accounts. With the present language, the definitions are equivalent in the sense that the partial functions generated by the two approaches are the same. In these circumstances, either account may be taken to define the programming language as a mathematical entity. However, in the way that the two are used in relationship to each other, the way they are treated in the literature [171, 173, 212], the operational account is in the driving seat: the denotational semantics must agree with the operational one. It would seem that the operational account is taken as the central definition of the language.

This definitional priority of operational semantics has its roots in the theory of computation where formal accounts of computability introduced formal languages with associated operational semantics. Turing machines and the lambda calculus are classic instances. In particular, the lambda calculus is given by an operational account that is very similar to the one given here except that it employs a small step semantics based on the following simple rules:

$$(\lambda x.t)s \Downarrow t[s/x] \qquad \frac{t \Downarrow s}{\lambda x.t \Downarrow \lambda x.s}$$

$$\frac{t \Downarrow s}{at \Downarrow as} \qquad \frac{t \Downarrow s}{ta \Downarrow sa}$$

The reduction is instigated in the first rule where the argument of the application (s) is substituted for the variable of abstraction (x) in the body (t). The rest allow the substitution to be extended to contexts. Such accounts are axiomatic accounts of the underlying notion of computation that the language embodies.

However, from a more traditional mathematical perspective, the operational approach is taken to be inadequate. To semantically specify a formal language, a *mathematical semantics* is required. The motivation for the denotational approach to semantics stems from an intuition that programs, and the objects they manipulate, are concrete realizations or implementations of abstract mathematical objects, i.e., programming languages refer to (or are notations for) abstract mathematical objects. In [213], this criticism is leveled at Landin's operational approach to the lambda calculus:

> We can apparently get quite a long way expounding the properties of a language with purely syntactic rules and transformations. ... One such language is the Lambda Calculus and, as we shall see, it can be presented solely as a formal system with syntactic conversion rules. ... But we must remember that when working like this all we are doing is manipulating symbols – we have no idea at all of what we are talking about. To solve any real problem, we must give some semantic interpretation. We must say, for example, these symbols represent the integers.

To give a set-theoretic semantics for the lambda calculus, we require a mathematical structure D such that terms of the calculus can be interpreted as elements of D in such a way that application in the calculus is interpreted by function application. This is made problematic by lambda terms such as xx. Given the standard interpretation of set-theoretic functions, if the second occurrence of x in xx has type D, and the whole term xx has type D, then the first occurrence must somehow have, a function space type of the form $D \to D$ for a class of functions that is closed under the operations of the calculus. Consequently, in some sense $[D \to D]$ must be contained in D. This is provided by some form of domain theory. The underlying mathematics is not trivial.

There is something quite odd about this position. The lambda calculus is defined by the rules of the calculus, not by any set-theoretic interpretation. The operational rules have definitional and epistemological priority. They are how the calculus is defined, and they determine how computation is carried out. Indeed, given the motivation for the calculus, the set-theoretic interpretation, as mathematically elegant as it is, seems inappropriate as a definitional apparatus. The calculus was developed to provide a mathematical theory of computable operations [36]. The uncovering of the appropriate notion of computability had epistemological intent. Set theory is the theory of infinite sets given in extension, and these do not seem to be the right referents for the operational notions of the calculus. In a similar manner, it would greatly distort the impact of Turing's account of computability if it was seen as fundamentally given through set-theoretic spectacles. The whole point of a fundamental definition of computability would be lost.

On this account, the denotational semantics has a purely mathematical role; it facilitates the exploration of the mathematical entity given by the operational account. Often properties of the whole theory are more easily established by exploiting the set-theoretic model. Programming languages implicitly provide theories of computation, and these are brought into being by operational semantics. Denotational definitions play a mathematical role.

10.5 Game-Theoretic Semantics

However, for this role of denotational semantics to work, we require that the denotational and operational accounts be in harmony. Unfortunately, this is not always so. This has led to the development of more fine-grained denotational approaches based on game theory [4, 5, 173]. The rise of distributed systems motivated the need for a model that allows interactions between components of a system. Abramsky and Jagadeesan [5] and Hyland and Ong [122] developed game-theoretic models for linear logic [86]. These models are intensional in nature: thus the usual completeness results, stating that provability of a formula is reflected in the model, were strengthened to *full completeness,* in which each proof is itself represented. This initiated intensional game-theoretic models for programming languages. These models provided denotational models in which the operational and denotational semantics agree. In particular, they solved the *full abstractness problem* for the programming language PCF [2], by giving the first syntax-independent fully abstract model. PCF is a higher-order functional programming language that is a fragment of any programming language with higher-order procedures. Full abstraction can be seen as a completeness property of denotational semantics that guarantees agreement between the notions of equality delivered by the two semantic definitions.

10.6 Programming Languages as Mathematical Theories

On the face of it, via their operational semantic definitions, programming languages are mathematical theories of computation [115]. However, although we might allow that elegant and simple theories such as the lambda calculus or our theory of programs for the simple WHILE language are worthy of mathematical status, we might still insist that this is not so for actual programming languages; what might hold for theoretical languages does not scale up. In particular, genuine mathematical theories must have some aesthetic qualities: they must have qualities such as elegance and the ability to be mathematically explored. And while it is possible to provide semantic definitions of the kind given for our toy language, for whole languages such definitions are normally not tractable theories. They are hard, if not impossible, to explore mathematically. They are often a complex mixture of notions and ideas that do not form any kind of tractable mathematical entity. Consequently, when provided, such semantic definitions are often complicated and unwieldy, and therefore of limited mathematical value. Given this, it is hard to argue that actual programming languages are genuine mathematical theories.

However, there is an observation that might be taken to soften this objection. And this involves the logical idea of a conservative extension. Suppose that we have constructed a semantic theory for a language. Suppose also that, in the sense of mathematical logic, we have shown that it is a conservative extension of a smaller tractable theory. Can we then claim that is also a mathematically acceptable theory? In other words, is a theory that is a conservative extension of a mathematical theory also a mathematical theory? A positive answer fits mathematical practice, where the mathematical exploration of a theory often results in the construction of conservative extensions. Indeed, the construction of these extensions is itself part of the exploration process of the base theory.

Programming languages admit of a similar distinction. While the whole language/theory may not have sufficient simplicity and elegance to be mathematically explored, it may nevertheless possess a conceptual core that can be. Such a core should support the whole language in the sense that the theory of the latter is a conservative extension of its core.

Unfortunately, there are further problems to overcome. No doubt there are some simple economies of syntax and theory that may be made for almost all languages. But it will generally be a nontrivial task to locate such mathematically acceptable cores for existing languages. Most programming languages have been designed with little mathematical input, and it would not be a simple task to locate an elegant core. But there is another route that involves the use of semantic techniques in language design. We shall explore this later.

Chapter 11
SEMANTICS AND IMPLEMENTATION

Programming language implementations are complex pieces of software and hardware that provide paradigm instances of computational systems. Actual implementations are often separated into phases involving syntax analysis, compilation, and interpretation [8], and involve layers of translation before a concrete representation is reached through direct interpretation.

In particular, a compiler is a computer program (or set of programs) that transforms source code written in a programming language (the source language) into another computer language (the target language). Compilers are programs, technical artifacts in their own right. As such, they have a function and structure. Lower down the implementation hierarchy, the structural description of the implementation must contain some account of the low-level mechanisms of the implementation, and ultimately its underlying logic machines. Many other artifacts arise naturally as parts and by-products of such implementations. Indeed, an actual implementation involves a complex weave of logically intricate technical artifacts that provide some of the most sophisticated examples of software and hardware systems.

Our first objective is to provide an account of programming language implementations as technical artifacts. Once this is in place, we shall discuss the main conceptual issue concerning implementation, namely, the view that an implementation provides a semantic interpretation of the language.

11.1 Compilers, Interpreters, and Virtual Machines

The production of physical devices from symbolic ones is seldom direct. In practice, the programs of high-level languages and their physical analogues are far apart. Indeed, the generation of the latter from the former usually involves compilation, i.e., a translation from programs in a source language into programs in a target language.

© Springer-Verlag GmbH Germany, part of Springer Nature 2018
R. Turner, *Computational Artifacts*, https://doi.org/10.1007/978-3-662-55565-1_11

As programs, compilers may be considered as technical artifacts in their own right. Functionally, a compiler must *correctly* translate between languages, where a correct translation must act in accord with the syntactic and semantic descriptions of the source and target languages. In particular, it must preserve the semantic content of the source language (L) relative to that of the target one. We illustrate matters with state-based languages. For simplicity, we assume that the structures of the underlying states of the two languages are the same. The correctness condition is then expressed by the following demand, where $[\![-]\!]$ indicates the semantic value of the programs in both languages:

$$\forall P : L \cdot \forall s : State \cdot [\![Compile(P)]\!]\, s = [\![P]\!]\, s.$$

In other words, if the source program with input state s outputs state $[\![P]\!]\, s$, then the compiled program, $Compile(P)$, follows suit. This is the heart of the functional description of a compiler as a technical artifact: its function is to correctly translate between the languages, and the core of correctness is the preservation of semantic content. In practice, the state machines of the two languages will be different, but the above captures the essence of the correctness criteria.

In contrast, the structural description of the compiler is the actual program that expresses the compilation. From this perspective, the design problem is the construction of a program that correctly translates between the two languages. If the symbolic program for the compiler is written in C, then its implementation must be for the language C, and the compilation stage must turn C programs into target code. With these notions of function and structure, compilers are technical artifacts with the following components:

Semantic definition $\rightarrow$ *Design* $\rightarrow$ Compiler code $\rightarrow$ *Implementation* $\rightarrow$ Physical Process

The function is the semantic criterion for the correctness of translation, the structure is the code of the compiler written, in a given programming language, and the artifact is the physical manifestation of the compiler code. Compilers might also be written in (possibly some subset of) the source language. There is an air of regress here in that not only is one language replacing another in the implementation process, but also the source language might be employed in the code of the compiler. But this is easily resolved, since at some stage compilation is replaced by interpretation, and ultimately by digital circuits. Of course, the actual physical code that is generated by the compiler is complex and hidden from the compiler writer. But, this is true for programs in general.

An interpreter is a computer program whose physical counterpart directly executes instructions. In principle, the source and target languages of a compiler may be at any level of abstraction, but eventually the compilation process results in assembly language instructions that refer directly to a machine's architecture. There is a very close correspondence between assembly language and machine code. Each assembly language is specific to a particular computer architecture, and is converted into executable machine code by another program called an assembler. Eventually, machine language instructions are translated into the language of a central processor, the language of ones and zeros. This language is then interpreted in terms of logic gates.

How is an interpreter to be interpreted as a technical artifact? What is its function and what is its structure? Conceptually, we can concentrate on the last step since the others are compilation processes. This step can be illustrated by reference to the instructions for any abstract machine, and the corresponding physical machine operations. We have already seen many instances of this in our account of logic machines.

Of course, there are more sophisticated options that bypass much of the translation phases. In particular, for efficiency reasons, the actual physical machine may be closer to the abstract mechanisms of the high-level source language. For example, Lisp machines were computers designed with Lisp as their machine code. They were hardwired in electronic circuits.

But, at whatever level interpretation occurs, at some point it involves a physical machine. This machine is itself a technical artifact with a function given by an abstract machine, and a structure given by the physical description of the machine.

More commonly, a language may also be implemented on a *virtual machine* where virtual machines [245, 248] are implemented in the underlying operating system. They offer a layer of abstraction that supports the language features in a self-contained environment. One of the authors to write clearly about the nature of virtual machines is Aaron Sloman. He distinguishes between two notions [208]:

> Two concepts of a "virtual machine" (VM), have been developed in the last half century to aid the theoretical understanding of computational systems and to solve engineering design problems. The first, Abstract virtual machine (AVM), refers to an abstract specification of a type of system, e.g. a Turing machine, a Universal Turing machine, the Intel pentium, or the virtual machine defined by a language or operating system, e.g. the Java VM, the Prolog VM, the SOAR VM, the Linux VM, etc. An AVM is an abstract object that can be studied mathematically, e.g. to show that one AVM can be modelled in another, or to investigate complexity issues. The second concept refers to a running instance of a VM (Running virtual machine, RVM), e.g. the chess RVM against which an individual is playing a game at a certain time, the editor RVM I am using to type these words, the Linux RVM running on my computer, the networked file-system RVM in our department and the internet – a massive RVM composed of many smaller RVMs, instantiating many different AVMs.

This distinguishes between abstract virtual machines and running ones, physical devices that actually run programs. The Java machine is an abstract machine that supports the intended semantics of Java. Its physical incarnation is the mechanism of physical computation. The point is that the term *virtual machine* is used to refer to both an abstract machine and the physical device. They provide yet another example of a computational artifact, with the abstract machine acting as the functional specification.

Other examples are given by the functional machines that provide the operational semantics for lambda expressions. These are high-level machines for functional languages such as Miranda and Haskell. SKI and graph reduction combinator machines [38] were once fashionable. Landin [141] introduced an abstract virtual machine version of such a functional machine known as the SECD (Stack, Environment, Control and Dump) machine. For Miranda programs, the abstract operations would be cashed out as a sequence of lambda reductions.

Computational artifacts are put together to form ever more complex ones. Indeed, as we have seen, the implementation process is itself very complex with a weave of interacting artifacts. Generally, computational artifacts may be put together in various ways to form new ones from old. Putting technical artifacts together via composition is one way. In particular, a programming language implementation may be seen as the composition of several technical artifacts: several compilers. Eventually compilers are composed with an interpreter, and compiler correctness is conjoined with the correctness of the machine code. In this way this complex software and hardware system that is a language implementation is itself a technical artifact.

An operating system is a technical artifact whose function is to control all the systems running on a computer. It is a software system that, among other things, performs memory management and controls the running processes. It manages the interactions within the computer and controls the access of each program to the computer's (CPU), memory, and storage. The Windows and Mac systems are common examples. Operating systems process other technical artifacts. They operate on compilers, run them, and generate other technical artifacts as output. The implementation artifact is called by the artifact that is the operating system, and is applied to the symbolic program.

It is operations such as these that utilize the levels of abstraction involved in specification and construction. This kind of complexity may well be present in the construction of standard technical artifacts. Undoubtedly, the manufacturing process for cars is also very elaborate. However, computational artifacts present new levels of logical complexity. The layers of abstraction, the rich interweaving of the technical artifacts involved, and the logical complexity of the languages and their elements make the technical artifacts of computer science some of the most complex artifacts ever constructed.

11.2 Compilation and Semantic Interpretation

With this background to the technical content of implementations, we are able to address the main conceptual issue of this chapter. There seems to be an informal belief among many practicing programmers that an implementation of a language provides its semantic interpretation. At least, in so far as they experiment with the implementation to figure out the meaning of the constructs, this reflects the practice of many programmers. Semantic definitions are taken to be irrelevant, or at best too difficult to take in.

Moreover, some philosophers appear to endorse the view that implementation provides the semantic definition of the language. For instance, Rapport argues that an implementation is always semantic interpretation [192, 193]. This would seem to imply that we can use an implementation as a semantic definition.

One interpretation would have it that semantics can be given in terms of compilation. In other words, the semantics of a language is given by translating it into another one. However, on the face of it, such an approach does not satisfy our normative requirements on semantic theories. In particular, by itself, it does not provide any criteria of correctness for the user or the implementer. Translations just pass the burden of meaning onto another language.

Of course, provided that the final interpreting language has been supplied with a semantic account, any single translation or series of translations may yield a semantic account of the source language, in which case, for each construct, this would involve the steps generated by the implementation that eventually unwind into a sequence of low-level machine code operations. Ultimately, this provides the semantics of the source language via the semantics of machine code. However, even though it ultimately supplies a normative semantic definition of the language, it defeats the whole purpose of high-level languages. Semantic accounts must reflect the high-level constructs of the language; they must operate at the level of the abstract virtual machines of the language.

There is a further complication. Any such translation involves not only the target language, but also the vehicle in which the actual translation is expressed – the metalanguage. So this too must be given a semantic interpretation. In other words, we have two languages for which we require a semantic account – the target language and the metalanguage. Unless both have semantic definitions, we do not have a normative semantics for the source language, one that provides the correctness criteria for the applications of the language. Consequently, it is hard to see how compilations satisfy the semantic demands of the programmer or the implementer. In the end, we have interpreted one uninterpreted language into another. Compilers or translators alone cannot provide definitional semantic accounts: a translation into a target language that itself has no semantic interpretation does not yield a definitional semantic interpretation of the source. So what form can a semantic interpretation in terms of implementation take if it is not an interpretation into another language?

11.3 Semantics and Interpretation

Interpreters provide the ultimate mechanism of implementation: if it is to be used, there has to be a physical interpretation of the programming language that runs programs. Can a physical interpretation, an interpreter for the language, furnish a semantics capable of fulfilling the semantic requirements laid out in Chapter 9? To illustrate matters, consider the simple assignment statement:

$$A := 13 + 74,$$

where A is a location in the state of a machine. What is its semantic interpretation? The physical interpretation might take the following form:

physical memory location A receives the value of physically computing 13 plus 74.

Consider for the moment the physical machine as a technical artifact. The function of the artifact must tell us what the machine and its physical components ought to do, not what they actually do. The problem with such a physical interpretation of the operations is that it does not distinguish between the physical structure of the machine and its functional requirements. Function concerns intention, not actuality.

If an actual physical machine is taken to contribute in any way to the meaning of the constructs of the language, then their meaning is dependent upon the contingencies of the physical device. For example, if these physical properties contribute to the meaning of any programming constructs, to determine what they mean, we must carry out a physical computation. In particular, the meaning of the simple assignment statement may well vary with the physical state of the device. So assignment does not mean assignment, but rather what the physical machine actually does when it simulates assignment. We are confusing the intended function with what the machine actually does. Indeed, if physical considerations enter into semantic interpretation, then even the evaluation of arithmetic expressions is subject to physical interference: addition does not mean addition, but rather what the physical machine actually does when it adds. Consequently, $13 + 74$ might be 86. This makes calculation an activity that is subject to causal interference.

Seemingly, physical machines in themselves cannot offer a definitional semantics; they cannot provide a guide to the user, and inform them what the correct use of an expression or construct is. This is not a contingent matter that can be entrusted to a physical device. This point is well made by Kripke in his discussion of rule following.

Actual machines can malfunction: through melting wires or slipping gears they may give the wrong answer. How is it determined when a malfunction occurs? By reference to the program of the machine, as intended by its designer, not simply by reference to the machine itself. Depending on the intent of the designer, any particular phenomenon may or may not count as a machine malfunction. A programmer with suitable intentions might even have intended to make use of the fact that wires melt or gears slip, so that a machine that is malfunctioning for me is behaving perfectly for him. Whether a machine ever malfunctions and, if so, when, is not a property of the machine itself as a physical object, but is well defined only in terms of its program, stipulated by its designer. Given the program, once again, the physical object is superfluous for the purpose of determining what function is meant [134].

A semantic account must provide an account that determines correct use, and implementation accounts based on physical implementations cannot do so. Reasoning about the effect of programming constructs must be subject to the same kind of rules that we employ for arithmetic. Axioms such as:

$$m + (n + 1) = (m + n) + 1,$$

are not subject to the contingent behavior of any physical device. For identical reasons, reasoning about the impact of assignment (or *while* loops or stacks) must be governed by similar axioms or rules such as the following:

$$(x := n; x := m) \equiv x := m.$$

The impact on the store of the left-hand right-hand sides of $\equiv$ is the same. On the assumption that the normative definitional requirement on semantics has to be satisfied by any adequate semantic theory, neither compilers nor interpreters fit the bill.

Much the same point can be made in terms of technical artifacts: to run together a semantic description with an implementation is to confuse *function* and *structure*. Despite this, some argue that implementation does contribute to semantics. Seemingly, this fits the duality perspective, according to which a program has both an abstract and a physical manifestation. Fetzer [70] observes that programs have a different *semantic significance* from theorems. In particular, he asserts the following.

> programs are supposed to possess a semantic significance that theorems seem to lack. For the sequences of lines that compose a program are intended to stand for operations and procedures that can be performed by a machine, whereas the sequences of lines that constitute a proof do not.

This insists that programs are intended to *stand for* operations that can be performed by a physical machine. Indeed they are, but this is not their semantic interpretation. However, the quote makes it clear that this referential aspect is part of the semantic significance of programs. Colburn [42] seems to agree, and makes matters a little clearer: he suggests that the simple assignment statement is semantically ambiguous between an abstract semantics and a physical interpretation such as the one given above. Semantic ambiguity suggests that both the physical and the abstract machines are involved in the interpretation of assignment. But this conflicts with our present semantic perspective, under which any physical interpretation with physical locations is seen as its implementation, which has to conform to the abstract interpretation i.e., it is correct or not relative to the abstract one. In other words, the physical implementation is subject to the abstract interpretation, and the meaning of the construct is given by the abstract account alone.

It may well be true that to use assignment one needs to know more than the given abstract interpretation. This will be so where the abstract account is not sufficient for the user to construct correct programs. This happens when the semantic definition does not cover the construct in sufficient detail for the programmer to ensure correctness relative to its specification. The user is then forced to employ the physical implementation to guess the intended meaning of the language components. But notice, it is the intended meaning that is required; it is not what the implementation actually does. It is still not the physical properties of the machine that play a semantic role. The result of a machine calculation once carried out may be assigned a normative function, but it will only be so assigned if it is believed to be the intended behavior. Presumably, the user will not take a normative stance towards a machine that returns the value 4 for all additions.

This does not entail that an abstract description of an implementation cannot serve as a semantic description. For example, the Java virtual machine is an abstract (virtual) machine on which an abstract interpreter for Java may be constructed. The description of the machine is a functional description of the artifact at some level of abstraction from the actual implementation. For example, the memory layout for run-time data, the garbage collection algorithm used, and any internal optimization of the Java virtual machine instructions (their translation into machine code) are not specified at the virtual machine level. Any Java application can be run only inside some concrete implementation of the abstract specification of the Java virtual machine. Such semantic accounts are functional descriptions of implementations. However, once taken as an abstract definition, any description of an implementation can function as a specification. This is closely related to a point made by Duhem [62]:

> When a physicist does an experiment, two very distinct representations of the instrument on which he is working fill his mind: one is the image of the concrete instrument that he manipulates in reality; the other is a schematic model of the same instrument, constructed with the aid of symbols supplied by theories; and it is on this ideal and symbolic instrument that he does his reasoning, and it is to it that he applies the laws and formulas of physics. A manometer, for example, is on the one hand, a series of glass tubes, solidly connected to one another filled with a very heavy metallic liquid called mercury and on the other by the perfect fluid in mechanics, and having at each point a certain density and temperature defined by a certain equation of compressibility and expansion.

Once treated as an abstract definition, any description of an interpreter can function as a semantic definition.

Rapport has it that implementation is always semantic interpretation [192, 193]. But does a concrete data type provide the semantic interpretation of an abstract one? Suppose we employ finite lists to provide an implementation of finite sets. Assume that the axioms for finite sets include axioms for the union operation on sets, for instance, the axiom that insists that the order in which the union operation is applied does not change the result, i.e., is commutative. Now suppose that the target language has lists as a data type. We may then implement finite sets as lists, where, for example, union is implemented as the concatenation (*) of lists:

$$Implement(a \cup b) = Implement(a) * Implement(b)$$

This interpretation, if correct, must preserve the union axioms for finite sets. Together these conditions provide the correctness criteria for the compiler – its functional demands. But, according to the semantic view of implementation, finite lists provide the semantic interpretation of finite sets. It would seem to follow that the axioms for lists fix the correctness of the axioms for finite sets. But this is the wrong way around. On the contrary, it is the axioms for the abstract type that must be preserved by the implementation. On the semantic view of implementation, the cart and the horse have been interchanged.

11.4 Programming Languages as Technical Artifacts

In fact, implementation semantics is really a special case of the causal theory of function. On the simple-minded causal theory, the function would be located in the physical implementation itself. This provides no definitional account of semantics, and no notion of correctness of use. We cannot distinguish between the implementation correctly or incorrectly implementing the semantic description of the language. Physical implementations in themselves cannot offer a definitional semantics; they cannot provide a guide to the user and inform them what the correct use of an expression or construct is. The semantics-as-implementation perspective, as an instance of the causal theory of function, inherits all the problems of the theory.

The alternative view sees programming language implementations as technical artifacts – where the syntax and semantics of a language act as the hub of its functional specification [232].

Chapter 12
SPECIFICATION LANGUAGES

To treat programming scientifically, it must be possible to specify the required properties of programs precisely. Formality is certainly not an end in itself. The importance of formal specifications must ultimately rest in their utility – in whether or not they are used to improve the quality of software or to reduce the cost of producing and maintaining software [120].

Specification is a heterogeneous enterprise with layers of complexity, generality and degrees of detail. In practice, specifications are expressed in a host of languages and formalisms. These range from the vernacular through to specialized specification languages. Some of them are graphical in content (e.g., portions of UML), and many others are based upon some logical notation. There are also algebraic approaches that employ algebraic or model-theoretic structures.

We shall focus on logical languages, since they are sufficient to illustrate the central semantic concerns. Moreover, we concentrate on languages employed for program specification. Our main objective concerns the logical and semantic foundations of these languages. What governs the choice of an underlying logic for these languages? Is classical logic appropriate or should some partial or constructive logic be employed? How should the semantics be given? Is set theory the right semantic medium? We shall examine these questions by reference to three systems: Typed Predicate Logic (TPL) [234, 237], Z [258], and VDM [127].

12.1 Typed Predicate Logic

We can highlight the conceptual issues with respect to a simple logical language (called TPL) [237] that is a typed version of predicate logic [19]. This will enable us to illustrate the definitional nature of specifications, and provide an account of the semantic issues associated with specification languages. It will also enable us to provide a smooth comparison between the foundational frameworks of mainstream specification languages such as Z and VDM.

© Springer-Verlag GmbH Germany, part of Springer Nature 2018
R. Turner, *Computational Artifacts*, https://doi.org/10.1007/978-3-662-55565-1_12

TPL in its minimal form [234] has the structure of predicate logic generated by atomic sentences $R(t_1, \ldots, t_n)$ (where equality between terms, $=$, is binary) and by negation ($\neg$), conjunction ($\wedge$), disjunction ($\vee$), and implication ($\Rightarrow$), but with quantifiers that are typed. The following BNF syntax summarizes matters.

$$\phi ::= R(t_1, \ldots t_n) | \neg\phi | \phi \wedge \phi | \phi \vee \phi | \phi \Rightarrow \phi | \exists x : T.\phi | \forall x : T.\phi$$

As usual, terms $(t, t_1, \ldots, t_n)$ are made up of variables, constants, and function symbols applied to terms. In the quantifier clauses, T is a type of the language. The types determine the ontology of the language; they fix the things that may be described or represented in the language. Moreover, in this general description of the language, like the notions of relation and term, the notion of type is a parameter. Different type systems allow specifications of different styles, and at different levels of abstraction. So, just as particular first-order theories in predicate logic introduce distinguished relation and function symbols, in typed predicate logic different type theories involve different collections of types.

Indeed, it is types that determine the level of abstraction or the application domain for which the language is appropriate. To illustrate consider the following language of types. The basic types include the types of numbers (N) and Booleans ($Bool$):

$$T ::= N | Bool | T \otimes T | Set(T).$$

The type constructors allow for the formation of Cartesian products of two types, the elements of which are ordered pairs (t_1, t_2) of objects from the given types. The type constructor $Set(T)$ has elements that are finite sets of objects of type T. This might be suitable for a specification which is to be employed for database languages.

For applications that support programming in the functional style, where, semantically, programs are mathematical functions, we might employ a system of types in which we replace finite sets with the type of operations from one type to another:

$$T ::= N | T \otimes T | T \Rightarrow T.$$

Types reflect the level of abstraction, and lower level types contain more implementation detail. For example, we might replace finite sets with lists:

$$T ::= N | Bool | T \otimes T | List(T).$$

Lists are a lower-level data type in which the order of occurrence is significant. We shall have more to say about levels of abstraction in Chapter 21. At this point we are concerned only with the use of predicate logic, in its typed form, for specification.

12.2 The Logic of TPL

Here we assume a natural-deduction sequent style, where judgments take the following two forms

$$\Gamma \vdash \phi \qquad \Gamma \vdash t : T$$

where ϕ is a single well-formed formula, T is a type and t is a term of the language, and Γ is a set of assumptions of two different kinds:

1. Type declarations of the form $x{:}T$, i.e., the declaration that the variable x has type T.
2. Well-formed formulas.

The first judgment insists that ϕ follows from the assumptions and the second, $t : T$, asserts that t is of type T.

The rules for the quantifiers are the standard ones taking the types into account. In particular, we assume the normal side conditions for the quantifier rules:

$$\frac{\Gamma \vdash \exists x : T \cdot \phi \quad \Gamma, x : T, \phi \vdash \eta}{\Gamma \vdash \eta} \qquad \frac{\Gamma \vdash \phi[t/x]}{\Gamma \vdash \exists x : T \cdot \phi}$$

$$\frac{\Gamma, x : T \vdash \phi}{\Gamma \vdash \forall x : T \cdot \phi} \qquad \frac{\Gamma \vdash \forall x : T \cdot \phi \quad \Gamma \vdash t : T}{\Gamma \vdash \phi[t/x]} \ .$$

There are two options for rules governing the propositional connectives, depending upon whether one adopts classical or intuitionistic logic. The classical interpretation follows the standard Tarski truth-conditional account. The propositional connectives are given their truth-conditional meanings and the quantifiers range over the types. The intuitionistic interpretation might take the form of a realizability interpretation where witnessing programs provide the means of realization. Whichever route is taken, in practice the construction of definitions relies more on the rules of the logic than on the semantic interpretations. The logic employed in practice has two functions: to establish required properties of any defined system, and to demonstrate program correctness. Do we need classical logic for this, or will intuitionistic logic suffice? This amounts to the question of whether proofs using the law of excluded middle are required in practice, for example in proving program termination. Indeed, VDM as a specification language employs yet another logic to deal with the partial functions that arise in arguing about termination and partiality [127].

The interpretation of the types is perhaps a more significant concern for the foundations of computer science. Traditionally, types would be interpreted as the sets of Zermelo-Fraenkel (ZF) set theory. Indeed, both the types and the operations of the language would be interpreted in set theory.

But should the semantics of the languages of computer science be given in set theory? Is it appropriate for languages that are intended to describe the function and structure of computational systems? Taken at face value, computer science deals with *types*, not sets. Is it appropriate that discrete notions of computer science should be modeled as infinite sets given in extension? And this question applies to both specification and programming languages.

TPL takes types as basic or fundamental, governed by rules of membership. For example, leaving out assumption sets, some of the rules for finite sets might be given along the following lines [237].

$$\mathbf{S_1} \quad \frac{}{\emptyset_T : Set(T)} \qquad\qquad \mathbf{S_2} \quad \frac{a : T \qquad b : Set(T)}{a \circledast_T b : Set(T)}$$

$$\mathbf{S_3} \quad \frac{\phi[\emptyset] \qquad \forall x : T \cdot \forall y : Set(T) \cdot \phi[y] \to \phi[x \circledast_T y]}{\forall x : Set(T) \cdot \phi[x]}.$$

The first demands that the empty set of type T is a set of type T, while the second allows a new element to be added to an existing finite set. The third is an induction principle for finite sets. This perspective takes types as fundamental being governed by these rules. As an approach, it stays close to the underlying computational intuitions, where conceptually computer science deals with data items and their types.

12.3 Definitions

In this family of languages, the central modeling or specification vehicle is the notion of a schema. Suites of such schemas model computational systems, where the language includes operations for building larger schemas from smaller ones. A schema holds two pieces of information: a declaration part that carries the type information of the variables, and a predicate part that is a well-formed formula of TPL. Essentially, a schema expresses a relation between the objects introduced in the declaration:

- a signature;
- a predicate.

In the signature, the variables of the specification are declared and associated with their types. In the predicate, the properties of the intended objects and their relationships are articulated. Symbolically, this gives the body of a specification the following two components:

$$x_1 : T_1, \ldots, x_n : T_n$$
$$\phi[x_1, \ldots, x_n].$$

Here $x_1, \ldots, x_n$ are variables and $T_1, \ldots, T_n$ are types. The declaration provides the declaration context for the proposition. Given this, we take definitions to have the following form.

$$R = [x_1 : T_1, \ldots, x_n : T_n | \phi]$$

We take schemas to introduce a new relation symbol R, whose type is given by the types in the declaration. More exactly, we take the above to introduce a new relation symbol R, whose logical content is governed by the following axiom:

$$\forall x_1 : T_1 \cdot \ldots \cdot \forall x_n : T_n \cdot \phi[x_1, \ldots, x_n] \leftrightarrow R(x_1, \ldots, x_n).$$

For example, using the last type system, the following is a specification of a sorting program:

$$Sort = [x : List[N], y : List[N] \,|\, Perm(x, y) \wedge Sorted(y)]\,.$$

This insists that the relationship between the input and output lists is such that the output is a sorted permutation of the input. This unpacks as follows.

$$\forall x : List[N] \cdot \forall y : List[N] \cdot Sort(x, y) \leftrightarrow Perm(x, y) \wedge Sorted(y)$$

Our next example is the composition of three functions. This employs the second type system.

$$Composition = [f : T \Rightarrow S, g : S \Rightarrow R, h : T \Rightarrow R \,|\, \forall x : T \cdot h(x) = g(fx))]\,.$$

These are definitions of new relations expressed in the language of TPL. Moreover, as definitions they act as conservative extensions of the underlying theory of types. With such a mathematical definition, a great deal of analysis is possible prior to any implementation. But the main point of introducing this language and logical framework is to provide a better insight into the logical and semantic issues that surround specification languages.

12.4 Z and Set Theory

The notion of a schema comes from the Z specification language [209, 258]. Z provides a calculus of schemas that allows for the formation of new schemas from old ones, and this allows for the structuring of large-scale specifications as suites of schemas. TPL can be seen as a computationally sanitized version of Z.

In particular, both the underlying logical framework and the interpretation is different. Z is cast within standard set theory, not the variable type theory of TPL. Although Z has types, everything is interpreted as in set theory. Indeed, Z is taken as a syntactic extension of ZF set theory. Presumably, set theory was chosen as a base theory because of its expressive power, and it is taken as a safe harbor for the interpretation of formal theories.

However, it less clear that it should play such a role for computer science. In the case of Z, one thing that is unclear is whether one ever needs the full power of ZF to model software systems. Here there is a distinction between types and sets. While one may wish for potentially infinite types such as the type of natural numbers or the type of finite sets, it is less than clear that the elements of any required types, the data items, need be other than discrete. The same is true of the type of operations. Even operations as intensional objects are given operationally via rules of inference, not semantically as infinite sets in extension.

The semantic interpretation of schemas is also different in Z and in TPL: in Z, they are interpreted not as the conservative addition of new relations, but as first order-theories. The declaration part introduces a collection of entities that are governed and fixed by the predicate part which now provides the axioms that govern them. Given that the underlying system is set theory and these theories are interpreted set-theoretically, their addition is still conservative.

12.5 VDM and Three-Valued Logic

On the face of it, specification in VDM [127] is close to the previous Z style but employs explicit preconditions. We first provide a reconstruction of VDM inside TPL. On our interpretation, VDM specifications take the following form:

$$R(x : T, y : S)$$
$$Pre : \phi[x]$$
$$Post : \psi[x, y].$$

Here x is the input, of type T, and y is the output, of type S, and ϕ and ψ are well-formed formulas of the logical language. ϕ is the precondition that lays out any conditions that the input must satisfy, and ψ is the postcondition, that expresses the required relationship between the input and output. More explicitly, such specifications may be understood semantically as a relation between the input and output types that may be unpacked in predicate calculus terms as follows:

$$\forall x : T \cdot \forall y : S \cdot \phi[x] \rightarrow (R(x, y) \leftrightarrow \psi[x, y])$$

If the precondition is satisfied, then the defined relation agrees with the postcondition. The following is a VDM-style specification of the square root function for real numbers:

$$SQRT(x : Real, y : Real)$$
$$Pre :: x \geq 0$$
$$Post :: y * y = x \wedge y \geq 0.$$

SQRT defines a relation between real numbers, where *Real* is the data type of real numbers, which may include negative reals; the precondition insists that the input is positive. According to the semantic account given above, the specification defines an abstract relation, $SQRT(x, y)$, that is determined as follows:

$$(C) \; \forall x : Real \cdot \forall y : Real \cdot x \geq 0 \rightarrow (SQRT(x, y) \leftrightarrow y * y = x \wedge y \geq 0)$$

By employing subtypes [234], this can be reduced to a style without preconditions as follows:

$$R = [x : \{x : T.\phi\} \mid \psi[x, y]].$$

Unfortunately, all this is not quite VDM. The latter does not introduce relations but partial functions. Consequently, the types of VDM take something like the following shape:

$$T ::= N \mid T \otimes T \mid T \leftrightarrow T$$

where $T \leftrightarrow S$ is the type of partial functions from type T to type S. The schema for the square root specification would then have the type $Real \leftrightarrow Real$. In itself this looks like an innocuous change, but it is not. It allows terms in the language that may not be defined, and so has an impact on the underlying logic: it must now deal with undefined terms. Subsequently, VDM takes the route of three-valued logic. Unfortunately, this has a devastating impact upon the whole logic where all propositions have three values. A better logic that only affects the quantifiers is a version of partial logic that takes into account nondenoting terms but maintains the basic logic. The other alternative is the one adopted by TPL, which is to take relations as primitive. Functions may be introduced as new primitive symbols when their domains and range types have been identified.

12.6 Types, Not Sets

We have argued that operational semantics is the defining vehicle for programming languages. Here we are adding an additional demand that applies to both programming and specification languages. This concerns the way types are defined and introduced. Types are intensional entities with decidable equality. If we wish to maintain the conception that computer science deals with computational and operational notions, then types should be taken as primitive, and defined by their operational rules such as the one given above for finite sets. This reflects how they are employed in practice as rules of type inference, and captures their primary function, which is to aid the construction of correct programs and systems. Interpreting them as sets has mathematical benefit, but it has little definitional value, and it certainly does not reflect how they are used in practice.

12.7 Expressive Power

One general motivation for specialized specification languages concerns the technical nature of specification. Like programming, the specification of software systems is a technical enterprise that requires precise and expressive tools. The worth of any resulting formal model of requirements largely depends upon the expressive power of the language, i.e., the representational facilities of the language such as its type structure. In addition, large-scale designs need structuring facilities that enable the construction of larger systems from smaller ones. These mechanisms have to be made explicit and their syntax clearly laid down. Operations such as *hiding* and the *conjunction* facilitate such structuring. It is obviously unsatisfactory to make up formalisms on the fly, as is sometimes done in practice.

Part IV
METHODOLOGY

According to Nigel Cross [46], design methodology is

> The study of principles, practices and procedures of design.

The aim of design methodology is to improve practice. It is normative, not descriptive. Science and design are taken to be significantly different. In particular, Simon [206], contends that the natural sciences are concerned with how things are, whereas design is concerned with how things ought to be. Indeed, this seems to be the standard perspective in engineering:

> Design in a major sense is the essence of engineering; it begins with the identification of a need and ends with a product or system in the hands of a user. It is primarily concerned with synthesis rather than the analysis which is central to engineering science. Design, above all else, distinguishes engineering from science [103].

In the philosophy of technology, design is characterized as the activity that moves us from functional demands to structural descriptions [138]. Programming, software design, and programming language design are all design activities. As such, they inherit the central questions and issues of the philosophy of design [179]. What is a well-designed program? What is a well-designed programming language? Which methods are employed to obtain good designs, and do they work?

Chapter 13
SOFTWARE SYSTEM METHODOLOGY

Software systems give computer science its practical potency. Physical library systems, banking systems, and management systems and structures existed before the advent of the computer. Libraries have always issued books, and banks have always taken our money. But, with the advent of the computer, these systems have been modeled in software. Indeed, the latter models are not just computational models of existing physical systems, but have been digitally enhanced and enriched.

The construction of these systems has given rise to many methods for their specification, design, and implementation [184, 214, 241]. These range from the simple *waterfall method*, where development is rigidly carried out in a specific order, taking in the *spiral* model, and ending in the least regimented *agile* model. All of them involve

- requirements elicitation;
- system design;
- implementation;
- testing and verification.

However, they differ in terms of the regimentation, rigidity, and intensity of their employment. In the waterfall method, the components are deployed once with intensity, and in a specific order. The spiral approach allows one to evaluate matters at the end of the first iteration, and if necessary repeat the whole intensive cycle. In contrast, the agile approach permits a more superficial analysis at each stage and recycles through the loop several times, while employing client feedback throughout.

In this chapter, we provide an overview of these software development methods. Our goal is provide some background to the central conceptual and methodological issues that underlie software development.

© Springer-Verlag GmbH Germany, part of Springer Nature 2018
R. Turner, *Computational Artifacts*, https://doi.org/10.1007/978-3-662-55565-1_13

13.1 The Waterfall Method

The waterfall approach [214] is the most systematic development method, in which the outcome of one phase acts as the input for the next, and each must be completed before the next can begin. The requirements analysis stage seeks to uncover the requirements of the system. This is dictated by the desires and needs of the clients. In the waterfall model, this is an intensive activity in which the demands of the client are carefully documented. Requirements are extracted from users, customers, and other stakeholders, where a great many techniques such as interviews, questionnaires, use cases and prototyping are employed. This may take some considerable time, but in this approach this must be completed before the next stage. In the system design phase, the requirement specifications are employed to develop the actual system design. The process of design is self-contained and driven only by the documented requirements, and is completed before the implementation stage, the stage where the system is implemented in software. Testing and verification are carried out at the end of the main process of software construction. In principle, everything is done in strict order. Unfortunately, if at any stage, there is a mistake, it is carried through to the next stage.

13.2 The Spiral Method

The spiral approach also involves intensive work at each stage, but allows for repeating the whole process, with a review at the end of each cycle. In other words, the spiral model permits several iterations of the waterfall model. While the cycles patch the rigidity problem that arises with the waterfall model, it does so at the overall cost of software development; this cost goes up with every new cycle. This can be addressed by being less intensive at each stage, i.e., a softer touch is applied to requirements analysis, design, and implementation. This is what motivates agile development.

13.3 The Agile Method

Instead of a sequential design process, the Agile methodology follows an incremental approach with several *sprints* around the waterfall circuit. The main idea is outlined in the *Manifesto for Agile Software Development* [112]. Although requirements analysis, system design, implementation, and testing all occur, they are employed in a much faster and less intensive way:

> The Agile movement is not anti-methodology, in fact many of us want to restore credibility to the word methodology. We want to restore a balance. We embrace modeling, but not in order to file some diagram in a dusty corporate repository. We embrace documentation, but not hundreds of pages of never-maintained and rarely-used tomes. We plan, but recognize the limits of planning in a turbulent environment.

In the agile approach, matters are much less rigorous, and much less documented. For instance, only a rudimentary set of requirements are gathered, in order to move rapidly on to the next phase, where requirements engineers develop a rudimentary project design and then begin to work on small modules. Mistakes may be addressed and customer feedback incorporated into the design before the next sprint is run. The aim is to produce working software as quickly as possible.

13.4 Methodology

How does the developer decide between these models of software development? The waterfall method demands in-depth analysis and documentation. The method demands a detailed and usually lengthy requirements analysis. The client knows what to expect; in particular, they will be given a clear estimation of the overall cost and duration of the project. All this will be documented and will inform any future projects of a similar nature. In other words, there is a potential contribution to *software development knowledge*. The manifesto [112] appears to be critical of this aspect.

However, since the whole product is tested only at the end, any problems are carried through the whole project. Moreover, a client's needs may be vague and may evolve through the project. Presumably, the waterfall model is best used when there is a clear set of requirements and speed is not paramount.

The spiral model does allow for reviews, but, since each iteration is essentially that of the waterfall method, it is expensive and time-consuming. Each iteration increases the time and cost of each project. Clearly, one could be less fastidious at each stage – giving us the agile approach.

The agile method allows for changes to be made throughout. At the end of each sprint, project priorities are evaluated with feedback from clients. Moreover, the testing at the end of each sprint ensures that problems are caught early, and permits revision of the requirements at the end of each sprint. However, there are some disadvantages. Badly formulated initial requirements, may cause major problems from which it is hard to recover, no matter how many sprints there are: fast and completely wrong is worse than slow and right. This is a tortoise-versus-hare choice. Presumably, the agile method is more suitable when rapid production is more important than the quality of the product.

These are very sketchy perspectives on the advantages and disadvantages of the common methods for software development. There are a good number of empirical studies on the various methods, but space does not permit a complete review of the literature. See [183] for an overview. However, the differences between them are not our focus here. From our perspective, they all have the same overall structure, and we shall be more concerned with the conceptual questions that stem from these various stages. Some issues involve the nature of requirements and specification. What kind of activities are they? How do we characterize the logical role of specification? What is a good software design? What is it for a design to be correct and successful?

Chapter 14
SPECIFICATION

In this chapter, we consider the nature and methodological concerns of requirements elicitation and specification. These activities raise a collection of overlapping conceptual questions and problems [235]. What is the relationship between requirements analysis and specification? How exact and complete do specifications need to be? What is the methodological role of specification?

The transparent case concerns a specification where what is being asked for is of a known kind. For example, if a sorting program is required, and the definition of sorting is in place, then little more needs to be said. The term *sorting* classifies a *kind* of program or algorithm. Of course, refinements are possible. One might wish the program to run with a certain degree of efficiency, for example, we might require average complexity $n \log n$ This would rule in programs based upon *quicksort* and *mergesort* but rule out *insertsort*. For mathematical notions such as sorting, such definitions are easy to come by. Matters are less clear in the case of software systems where the first phase involves extracting or eliciting some reasonably exact notion of what is required.

14.1 Requirements Analysis as Modeling

In general, the function of a software system is complex. A function or purpose may have many aspects or parts that need to be spelled out individually, as well as collectively. In particular, complex software specifications have structure [154, 235]. Moreover, the process of determining this structure is fraught with difficulty. Techniques such as brainstorming, focus groups, prototyping, and surveys aid the elicitation of requirements [37]. Brainstorming produces an initial set of ideas. Prototyping is employed to extract requirements from nontechnical users, and focus groups and surveys facilitate the gathering of data from a wide variety of stakeholders. The intensity and scheduling of this activity vary from method to method, but after some work we might arrive at an initial set of requirements for a library system:

© Springer-Verlag GmbH Germany, part of Springer Nature 2018
R. Turner, *Computational Artifacts*, https://doi.org/10.1007/978-3-662-55565-1_14

A town requires a new library system. It must support the cataloging of new books, and the borrowing and returning of books by readers. It must have a system for acquiring books from vendors, and it must support all the normal security and financial arrangements concerning the borrowing and returning of books.

The techniques of elicitation listed above are techniques for forming a summary of the client's requirements for the proposed system. But what is the logical status and function of the requirements? We suggest that requirements elicitation is an attempt to build a *model* of the client's requirements. And by this we mean a model that captures the empirical demands of the stakeholders. It is a model that is subject to revision by reference to the clients' and end-users' demands. Unfortunately, clients may be vague about what they require; they may include unnecessary information while excluding necessary detail. The client may demand further functionality. So, as a model of requirements, this model stands or falls depending upon their requests and demands, and consequently may have to be revised. Indeed, the process of elicitation may involve several cycles or revisions until some stability is achieved. Whether this process takes place once and for all at the beginning of the project or as a continual process throughout the project is not our focus. We are concerned with its logical status, and its methodological role. And these are the same no matter how intense the elicitation stage, and where and when revision occurs.

Under any regime, the requirements analysis results in a *model* of the client's demands, and the process of evaluating it against those demands is an empirical activity governed by the demands of the client.

14.2 Definition

Once in place, these requirements form an abstract system that can stand on its own. At this point an intentional shift occurs. It is, as Wittgenstein puts it, "hardened into a rule" [256]. What starts as an empirical proposition, subject to revision, is hardened into a rule. This is how Wittgenstein sees the empirical investigation stage of mathematical activity when it crosses the border, and is baptized as a mathematical definition. In requirements capture something similar happens: we are experimenting in order to extract the client's requirements. However, once stability has occurred, we may treat the result as a stipulative definition [98] of a system.

A court can define a new legal notion such as internet fraud or cyberspace arson. Such notions often evolve over time. Legal definition is a complex, evolving process subject to empirical input and conceptual analysis. Likewise, definitions in mathematics evolve and change over time. Mathematics progresses by generalizing notions to more general settings, while at the same time preserving their application to the older ones. Finally, philosophers seek the correct definition of natural kinds or aim at isolating the real and nominal essence of a concept. The objects of these various forms of definition are quite different: they define legal notions, mathematical structures, and philosophical concepts. They also have quite different functions. The lawyers seek definitions to form the basis of laws. Definitions in mathematics play a variety of roles, but one central one is the stipulation of mathematical notions and structures. For example, the definition of a group determines an abstract structure and allows the normal mathematical enterprise of investigation to begin. Finally, philosophers seek conceptual analysis. Goodman introduced the concept *grue* to be a property of an object that makes it appear green if observed before some future time t, and blue if observed afterward. This was introduced to reveal a further puzzle, other than Hume's, with the standard account of induction.

Clearly, the activity of definition serves a range of purposes or functions. But there is always a process of investigation and uncovering of the required concept. In the mathematical case definitions evolve and change by a process of conceptual analysis and empirical evaluation [142]. Eventually, matters settle, and a definition is adopted by the relevant mathematical community. In a similar, if not identical manner, requirements elicitation settles on a definition of the required system. As definitions, they have a neutral role as the stipulation of a system. In our example, we stipulated the structure of a library system. Definitions form the underlying scaffolding of specification, and constitute a further role for definitions.

14.3 Intention

In themselves, definitions introduce abstract entities. They are constituted by the demands unpacked by the requirements, but once in place they stand alone as definitions that may be explored for their consequences. At this point, a further intentional shift occurs. Definitions are put to work; they are employed as functional specifications of actual systems. As such, they tell us what the system is supposed to do. Definitions are taken to point beyond themselves to the construction of an artifact. Acting as a specification, the definition determines whether or not any constructed system has been correctly built. If one asks whether the device works, it is the definition functioning as a specification that tells us whether it does. Without it, the question would be moot. Subsequently, any structural description of the artifact must conform to it. If I specify addition and the programmer provides a program that computes multiplication, it does not satisfy the specification. In order to repair matters, we may need to investigate how the mistake was made. However, the methodological role of specification is not to explain how, but why it is a mistake. Indeed, throughout the software construction process, the methodological role of specification is always the same: it provides a criterion for correctness and malfunction. This normative role is taken to be part of any general theory of function [139]. This is the perspective argued for in [235].

According to this analysis, there are two components to specification:

- definition;
- intention.

The propositional content of the specification is the definition. On the other hand, the intentional aspect enters when an agent takes the definition as a specification of a program or system. So, there is a distinction between the propositional content of a definition and its employment as a specification. It is the intentional act of giving it governance over the construction of an artifact that turns the definition into a specification. It is this that sanctions the employment of the definition as the vehicle of correctness.

Notice that a change occurs when the (neutral) definition changes its role from model to specification. As a model, it is intended to reflect and capture the demands of the client. As such it is revised until some stability occurs. But, once fixed, it changes its purpose to be a specification of the system. Now, if the constructed system does not satisfy it, it is the constructed system that is revised. We shall refer to this as an instance of *intentional shift*, and have more to say about the intentional aspect of correctness when we consider the notions of correctness that govern computational artifacts.

It should go without saying that any such analysis of specification is an idealization. It is a rational reconstruction of practice. In particular, the relationship between specification and artifact may be somewhat flexible in practice. Moreover, even if logically coherent, the artifact may still be impossible to construct for practical reasons: physical constraints, and time and cost limitations may all prohibit construction. In these cases, the current specification will have to be revised. This is a part of the caveats that govern specifications.

14.4 Intentional Stance

We can illustrate the intentional aspect of specification in a more general setting. Consider the standard semantics for the language of first-order logic. This provides a definition of truth for the language. However, it may be given governance over the construction of a proof system that could act as the structural description of an implemented theorem-proving system. In the background is the assumption that the semantic interpretation of predicate logic is a definitive account of the semantics of the language. Our intention is to construct and establish the correctness of an independent set of proof rules. Consequently, we will be engaged in a mathematical activity where success is traditionally taken as the soundness and completeness of the rules. Soundness establishes the legitimacy of the rules, and completeness demonstrates that we have not missed any. When there is disagreement between the proof theory and the semantics, we blame the proof theory. We change the rules to gain soundness and completeness.

Conversely, if we take the proof theory to have definitional priority, the task is to construct a semantic theory that is sound and complete with respect to it. Indeed, this better fits the picture with intuitionistic logic and Heyting's axiomatization, where it proved difficult to locate a satisfactory semantic interpretation. In justifying the semantics, we would still be engaged in a mathematical activity, and we would still attempt to prove soundness and completeness. However, soundness now demonstrates that the semantics is a correct reflection of the rules, and completeness demonstrates that the semantics does not sanction illegitimate ones. The proof theory now has governance. In particular, if there is a mismatch, if there is disagreement of either kind, we now blame the semantics.

Notice that the mathematical results are independent of the intentions involved: we would prove the same mathematical results whether we took the semantics or the proof theory to have priority. What is required to demonstrate the correctness of a semantic interpretation is the same as that required to demonstrate the correctness of any proof theory. The correctness criteria are identical. However, the *intentional stance* is different. We are engaged in the same proofs but with different intentions: what I take to have definitional priority determines the interpretation I give to soundness and completeness. If I take the semantics as having priority, then I must know that I am evaluating the formal rules. On the other hand, if I take the proof theory to have priority, then I must know that I am evaluating the semantic account.

The lambda calculus and its semantics provide a different example that is closer to the present application. We might be engaged in demonstrating that the rules of reduction are sound under some domain-theoretic model [3]. The soundness of the reduction rules counts as success; we do not have completeness under the domain-theoretic interpretation. The originators of this semantic perspective employed the rules to measure the success of the model. Clearly, it would only be a model if the rules were sound. In this sense, the proof theory of the calculus has priority over the semantics, in which case, the rules are a measure of the legitimacy of the domain-theoretic interpretation. A similar perspective operates with the operational semantics of all programming languages. We have argued that the operational semantic account has definitional priority. The denotational or set-theoretic semantics must agree with it. Ideally, it should be fully abstract. When full abstraction is not achieved, we massage the game-theoretic account to fit.

The *intentional stance*, what has governance over what, matters. In particular, it determines what we do when things go awry. It is at the heart of specification. It determines what is *function* and what is *structure*. It overrides the simple intuitions about the *what* and the *how*.

14.5 Precision and Information

How precise do specifications have to be? How much information about the required artifact needs to be given in a specification? To unpack this a little more we need to draw a distinction between *precision* and *information*. Dijkstra puts matters as follows:

> the purpose of abstraction is not to be vague, but to create a new semantic level in which one can be absolutely precise [54].

The objective of specification is to lay out the function of a program or system in a precise manner. The specification contains information about what is to be constructed, but it is at a higher level of abstraction than the information contained in the next level down. However, it may still be precise. That is, specifications may be precise even when they are expressed using very abstract notions; they need only be precise at the level of abstraction at which they operate.

This is best illustrated with a simple example. Consider finite sets and lists. The sets themselves might be given using axioms and rules such as those laid out in the discussion of TPL. For computational purposes, such axioms characterize the notion of a finite set. Given such axioms and rules we can describe a system using finite sets in a precise mathematical way. For example, consider the *Update* operation used to interpret assignment. There is no ambiguity about what the operation does at the level of sets. A state is a mathematical function from a finite set of *locations* L to a finite set of numerical *values* V. *Update* is defined in terms of deleting an element from a set of ordered pairs and adding a new one. Nothing more needs to be said at the level of abstraction characterized by finite sets.

More detail, a move towards an implementation, would be to define the update operation on lists. Here it would be natural and necessary to include implementation details about exactly how the list of location-value pairs was updated. We might assume that the new location-value pair is simply added to the list, and a program called a garbage collector later removes the old pair. Or we might remove the old one, and place the new one in the same place as the old pair. If a new location is introduced and assigned a value, we need to spell out how the list is updated – do we add the new location at the head of the list or somewhere else? With lists we naturally have more decisions to take, more implementation details to spell out. However, from the perspective of the levels of abstraction involved (sets or lists), while one contains more information, both are equally precise. Different levels of abstraction do not demand different levels of precision, but different levels of information. The notion of a finite set and the union operation that enables the addition of new elements is mathematically precise. So is the insert operation on lists. The latter is not more precise, but closer to the actual physical implementation: it contains more implementation information, more information about the structure of the device. This is close to the perspective of Floridi [72] (see also [82]).

In some forms of programming, a form of specification occurs in the programming language itself [39]. It does so via the type system of the interface of modules and classes. Interface specifications in some form or other are common in practice. One of the benefits is that specification and programming (implementation) take place in the same language, and may be automatically checked for type consistency. However, there are limitations to such an approach. For example, type specifications would not distinguish between set intersection and set union. Such specifications are not precise at the level of abstraction at which they are operating. Programming languages do not offer the required expressive power to impose further conditions without resorting to actual implementation. The best they can do is to use natural language side conditions to convey more information about the operations.

The arguments for a formal language such as Z are built upon this distinction between precision and levels of information: specification is a technical matter that requires precision at the appropriate level of abstraction.

Chapter 15
THE PHILOSOPHY OF DESIGN

Design [16] is everywhere in computer science. It occurs in the design of software both at the level of overall system design [9] and in the design of individual programs [93]. For simple programs, the activity of design may be implicit. But design decisions are, or have been, taken even in the most straightforward cases. For example, any programmer who chooses *quicksort* instead of *mergesort* has taken a design decision. More demandingly, it occurs when the programmer moves from a given specification to a program that she has never constructed before. For large systems, design decisions about the overall structure of a system, its parts, and how its functional demands are to be met take place long before any code is cut.

In this chapter, we introduce programming and software design from the perspective of the philosophy of design. Parson's book *The Philosophy of Design* [179] draws a distinction between the *theory of design* and the *philosophy of design*. He suggests that the main focus of theory is the practice of design, its rational reconstruction and its improvement. By contrast, the *philosophy of design* should:

> Examine design, and its specific aims and problems, in light of fundamental questions that philosophy examines; questions about knowledge, ethics, aesthetics and the nature of reality.

One fundamental question concerns the nature of design:

- What is a design, what kinds of things are designs, and what is a good one and how do we judge?

This raises basic metaphysical, epistemological, and aesthetic questions about design. What kind of thing is a design? What makes a design a good or successful one? What is it for a design to satisfy its functional specification? How is an explanation of a function in terms of a structure possible? Of course, these questions cannot be completely separated from the methodological ones:

- Which methods are used in practice to obtain good designs? How do we evaluate them?

The design process and the design product are intimately tied to each other. However, the analysis of *goodness* for design must come before any attempt to improve design practice [139] – which is

© Springer-Verlag GmbH Germany, part of Springer Nature 2018
R. Turner, *Computational Artifacts*, https://doi.org/10.1007/978-3-662-55565-1_15

generally taken to be the main goal of theory. As Kroes [138] argues, knowing what kind of thing a design is, knowing what a good one is, and how the nature and process of design are related, is a prerequisite for any kind of improvement. These questions apply to design in general, but we are concerned with their application to computer science and, in particular, to programming, software development, and programming languages.

15.1 Correctness

Correctness is a necessary ingredient of good design. Above all else, a program or software system must satisfy its functional specification. On the face of it, there are various notions of correctness that correspond to the relationships between function and structure on the one hand, and between structure and artifact on the other. The first notion seems to be formal or mathematical in nature, since it involves the connection between two formal notions. The second concerns the implementation, and in the case of programs this eventually involves the correctness of physical devices, and this is not a mathematical affair. These notions raise several philosophical concerns that we shall consider in the last part of the book, devoted to epistemological issues in computer science. Here we are concerned with notions of *goodness* that are taken to aid and support the construction, recognition, and verification of correct programs and software systems.

15.2 Simplicity

There are a whole host of things that are said about good programs and software designs. They are said to be elegant, transparent, simple, generic, uniform, and efficient. They are said to be easy to grasp and understandable by others. In the next few chapters we shall consider these notions in more detail, and attempt to unpack their nature and motivations. However, the most mentioned property seems to be *simplicity* [198]. In regard to correctness, Tony Hoare puts it as follows:

> The price of reliability is the pursuit of the utmost simplicity [117].

Simple designs are easier to grasp and easier to investigate. But what is simplicity in design? How is it measured? Exactly how does simplicity aid correctness? There are other notions that are often associated with simplicity. Uniform solutions are taken to be better than those that are piecemeal or require lots of disparate cases. Software that is more generic is also taken to be superior. These traits of design are often encouraged through the design of the languages themselves. We shall take up the issue of program and software simplicity in the next chapter, reserving a discussion of how the notion is applied to programming languages for a later one.

15.3 Modularity

What are the methods employed in computer science to obtain simple designs? Generally, development methods are governed by *software design principles*, principles that resemble those of the central engineering disciplines. One such is the principle of *nonrepetition* that demands that we reuse designs, while another entices us to minimize up-front design: only design what is necessary. But by far the most important is *modularization*, the process of breaking up complex problems into smaller, *simpler* ones.

One older version is encapsulated in structured and modular programming. In particular, the latter advocates that a complex system should be organized as a set of distinct components that can be developed independently and then plugged together. Certain languages, such as Modula, Ada, and Pascal, are designed to enforce such program structure. Modular programming frequently employs a top-down design model in which developers map out the overall program structure into separate subsections. A defined function or set of similar functions is coded in a separate module. However, almost any language can use structured programming techniques. Most modern procedural languages include features that encourage some form of modularity. Floyd's Turing Award lecture [71] contains some further reflections on structured programming, and its impact upon software development.

In the contemporary scene, modularization is governed by many auxiliary notions, aphorisms and principles. Encapsulation tries to ensure that programs are contained units with transparent and self-contained functional specifications. One is advised to avoid *leaky* modules that refer to lower levels of abstraction. The *separation of concerns* principle advises that applications be divided into distinct modules with as little overlap in functionality as possible, while the *single-responsibility* principle insists that each module should be responsible for a single aspect of overall functionality. The principle of *least knowledge* demands that a module should not know about internal details of other components or objects, and *self-containment* means that individual modules may be designed independently from each other.

These are principles of *modularization*. But how is modularization to take place: what are the principles that govern the process? An examination of these notions and issues will form the basis of Chapter 17 on modularity, modules, and modularization.

15.4 Formal Methods

The design principles we have just mentioned do not prescribe further how detailed development must take place. They are software design principles that are aimed mainly at large-scale software development. When it comes to the construction of individual modules or programs, other techniques have been advocated, and these include formal methods. Design principles such as those associated with modularity may operate at the top level of system design. They are global principles that dictate how we break up the complexity of large software systems. As such they are not obviously at odds with formal methods. Indeed, specification languages such as Z have structuring facilities that aid modularization.

However, in practice, formal methods are rarely used. Why? Some suggest that they extend the time period of the software cycle, and others that engineers do not have the correct mathematical background or that formal methods do not fit into the software cycle in a smooth way.

The use of formal methods is controversial. In this regard, we need to distinguish between the employment of formal languages as vehicles of specification and the process of formally deriving programs from specifications. Indeed, formal languages are commonly used at the specification and design stages, even if the formal languages employed are programming languages – or some features of them such as interfaces, classes and modules. Instead, the controversy is more firmly located in the construction of formal correctness proofs, and in the use of formal techniques for the derivation of programs from specifications. On the positive side it is said, and even argued, that formal programming aids the construction or derivation of correct programs. On the other hand, there is a widespread belief that formal methods, even when restricted to the coding stage, slow down software development. So, there appears to be a conflict between speed of development and correctness. We shall consider this controversy in Chapter 18.

Chapter 16
SIMPLICITY

One of the cornerstones of program and software design centers on the opposing notions of *simplicity* and *complexity*. Simplicity is said to aid transparency and reliability in design. So say professional computer scientists of all flavors. All put simplicity at the core of good and successful design. Hoare [117], Dijkstra [54], and Wirth [252] are prominent examples:

> There are two ways of constructing a software design: one way is to make it so simple that there are obviously no deficiencies, and the other way is to make it so complicated that there are no obvious deficiencies. The first method is far more difficult. It demands the same skill, devotion, insight, and even inspiration as the discovery of the simple physical laws which underlie the complex phenomena of nature [117].

> Simplifications have had a much greater long–range scientific impact than individual feats of ingenuity. The opportunity for simplification is very encouraging, because in all examples that come to mind the simple and elegant systems tend to be easier and faster to design and get right, more efficient in execution, and much more reliable than the more contrived contraptions that have to be debugged into some degree of acceptability [58].

> Increasingly, people seem to misinterpret complexity as sophistication, which is baffling – the incomprehensible should cause suspicion rather than admiration. Possibly this trend results from a mistaken belief that using a somewhat mysterious device confers an aura of power on the user [252].

Hoare suggests that it increases reliability; it helps to ensure correctness. Dijkstra insists that simple and elegant systems are easier and quicker to design, and get right. Wirth argues against complexity, and is puzzled why people associate it with sophistication and good design. The literature and the internet are full of quotes from practicing programmers and academics advocating simplicity in design. While there is a range of things being claimed for simplicity, all seem to agree that it is virtuous. However, from these quotes alone we do not gain much insight into what software simplicity is. Nor do we glean exactly how it contributes to good design. In this chapter, we shall address some of these issues. In particular, we shall focus on the following:

1. How is simplicity defined for software?

© Springer-Verlag GmbH Germany, part of Springer Nature 2018
R. Turner, *Computational Artifacts*, https://doi.org/10.1007/978-3-662-55565-1_16

2. What is the role of simplicity principles in software development?

3. Is there a rational way to justify such principles?

While we are concerned with simplicity in computer science, we can gain some help from the analysis of simplicity in other areas and, more specifically, from the philosophy of science. A good amount of work has been done on the analysis of simplicity as it applies to scientific theories. In particular, Baker [17] draws a distinction between two different notions of simplicity

- syntactic simplicity;
- ontological simplicity.

Baker refers to these two guises of simplicity as *elegance* and *parsimony*, respectively. As we shall see, versions of these notions are identifiable in the writings of computer scientists, so this distinction provides a good starting point for our analysis of software simplicity. However, these terms do not mean the same thing for programs as they do for scientific theories.

Applied to scientific theories, syntactic simplicity has something to do with the number and complexity of hypotheses in a theory, and the ontological variety refers to the number and complexity of things postulated by the theory. For example, Newtonian physics employs notions such as *mass* and *force*. However, in the computational literature, syntactic simplicity or elegance is harder to pin down. It is not the purely syntactic properties of the string of symbols that make up a given program. Nor is it straightforwardly related to the computational complexity of the program, i.e., its time and space complexities. It is something more ethereal; it is more akin to elegance applied to mathematical proofs. On the other hand, ontological simplicity for programs, like its scientific counterpart, seems to come down to a principle of parsimony, but not an ontological one. The aphorism here seems to be *solve the problem and no more*. This links the design with its specification, and insists that we do not go beyond the demands of the functional specification. However, on the face of it, parsimony and elegance might pull in different directions – since solving more than the required problem might be more elegant. This parallels what happens in mathematics, where generalizations may provide more elegant solutions.

Indeed, and this is a topic we shall take up later, elegant programs lead to elegant proofs of correctness. And one characteristic of elegant proofs is their explanatory power: they explain why the programs meet their specifications; they explain how they solve the problem.

16.1 Elegance

The writings of many computer scientists tie simplicity to some notion of beauty or elegance. One single source of such writing is the book *Beautiful Code* [174], where professional programmers explain how various notions of beauty motivate and guide them. Indeed, beauty seems to be mentioned everywhere design is present:

> Ugly programs are like ugly suspension bridges: they are much more liable to collapse than pretty ones, because the way humans (especially engineer–humans) perceive beauty is intimately related to our ability to process and understand complexity. A language that makes it hard to write elegant code makes it hard to write good code. Beauty is more important in computing than anywhere else in technology because software is so complicated. Beauty is the ultimate defense against complexity [194].

> When I am working on a problem, I never think about beauty. I think only of how to solve the problem. But when I have finished, if the solution is not beautiful, I know it is wrong [28].

Designers of all varieties appear to agree that ugliness is to be avoided, and somehow elegance guards against, and aids our ability to deal with, complexity [85]. But are we any further forward: have we not just replaced one unknown with another? Fortunately, we can push matters a bit more. While it is not easy to say what elegance for programs amounts to, the analogy with mathematical proofs is instructive: mathematicians talk about proofs being elegant and, in particular, being graspable or able to be taken in. We shall have more to say about elegant proofs in the next part of the book where we discuss correctness proofs for programs, but we can already note that much the same has been said of programs:

> If you cannot grok the overall structure of a program while taking a shower, e.g., with no external memory aids, you are not ready to code it [180].

One assumes that the shower is not directly relevant: it could be on a bus or in the bath. To make matters more concrete, consider the following program that sorts a list of numbers into numerical order:

```
function quicksort(list)
    less, equal, greater := three empty lists
    if length(list) > 1
        pivot := select any element of list
        for each x in list
            if x < pivot then add x to less
            if x = pivot then add x to equal
            if x > pivot then add x to greater
        quicksort(less)
        quicksort(greater)
list := concatenate(less, equal, greater)
```

It is easy to grasp the whole program as a whole i.e., the way the underlying algorithm works can be taken in. Indeed, how it works can be expressed in English as follows.

```
Choose any element of the array to be the pivot.
Divide all other elements (except the pivot)
into two partitions.
All elements less than the pivot must be in
the first partition.
All elements greater than the pivot must be in
the second partition.
Use the whole procedure to sort both partitions.
Join the first sorted partition,
the pivot, and the second sorted partition.
```

So, one notion of elegance concerns the graspability of programs, and seems to be associated with the clarity of the underlying algorithm.

A related property concerns correctness: elegant programs more transparently satisfy their specifications. For instance, it requires little work to show that the above program sorts. A simple inductive argument on the length of lists suffices. If the length is one, it is already sorted. If not, we split the list, and assuming inductively that its constructed sublists (less and greater) are sorted, it is easy to see that the program returns a sorted list.

There is also a close relationship between elegance and explanation. For example, Quicksort wears its explanation on its sleeve; transparency goes hand in hand with explanatory power. Asked why it works, something like a sketch of the above proof would suffice: the way it works explains how it works. Elegant programs give rise to proofs that have high explanatory power.

Another significant way in which elegance might be achieved is through uniformity.

> The key to performance is elegance, not battalions of special cases. The terrible temptation to tweak should be resisted unless the payoff is really noticeable [21].

Here elegance is identified with the absence of special cases. Uniformity encourages a striving for consistency in design. It urges the designer to aim at more general solutions: solutions that consist of a long list of case statements are not elegant; they are brute force solutions. It insists on solutions that involve few special cases. For example, a program aimed at computing the square roots of natural numbers that takes the form of a table as below, a set of case statements, would not be seen as elegant. Uniform solutions are taken to be superior in that the problem has been more abstractly understood, and the solution is expressed at a higher level of abstraction:

```
Case x=1 then y=1
Case x=2 then y=1.41421356237
Case x=3 then y=1.73205080757
```

Case x=4 then y=2
Case x=5 then y=2.2360679775

Programs at the right level of abstraction isolate the core of the problem from surrounding noise, and implementation details. Of course, this form of elegance is aided by the facilities of the language. Moreover, uniform solutions have greater explanatory power.

A stronger form of uniformity involves types: elegant solutions are generic. For example, a parametrized or polymorphic solution is more uniform than individual solutions for special instances of a problem involving different types. For instance, the following is a polymorphic version of the quicksort program that provides a uniform solution for all types:

```
quicksort :: Ord a => [a] -> [a]
quicksort [] = []
quicksort (p:xs) =
(quicksort lesser)++[p]++(quicksort greater)
where
lesser = filter (< p) xs
greater = filter (>= p)
```

Here each type instance generates a sorting function for the input type. Although it is hard to pin down exactly what computer scientists have in mind by *elegance*, the following seem to be significant features. The program should be:

- transparent;
- correct;
- explanatory;
- uniform;
- generic.

The emphasis here is on comprehension and correctness. This offers an epistemic justification for elegance in programming: elegant programs can be more easily comprehended and seen to be correct. See the recent work of Hill [113] for a different take on elegance in programming.

16.2 Parsimony

While not distinct from elegance, ontological simplicity for programs, like its scientific counterpart, seems to come down to a principle of parsimony [163]. Many professional computer scientists see parsimony as a characteristic of simplicity, and thereby of good design. But in computer science it is interpreted not in terms of things postulated but in terms of a notion of *minimality* i.e., the demand that a solution does not go beyond what is required. In a more general design setting, John Pawson in *Minimum* [181], puts it as follows:

> The minimum could be defined as the perfection that an artifact achieves when it is no longer possible to improve it by subtraction. This is the quality that an object has when every component, every detail, and every junction has been reduced or condensed to the essentials. It is the result of the omission of the inessentials.

Other writers on design seem to agree on this notion of parsimony: we should do no more than what is necessary:

> The ability to simplify means to eliminate the unnecessary so that the necessary may speak [119].

> The inherent complexity of a software system is related to the problem it is trying to solve. The actual complexity is related to the size and structure of the software system as actually built. The difference is a measure of the inability to match the solution to the problem [108].

This is Occam's razor applied to programs and software. We should write programs and software that solve the problem, fulfill the specification, but no more. It is rational to prefer the more parsimonious solution since it reduces the possibility of errors. In software development, designers often over engineer. This leads to unnecessary complexity, inefficiency, and, worse, incorrectness. A parsimonious solution to a problem should match the specification of the problem; it should not solve more than that required by the specification.

What does it mean to say that one program is more parsimonious than another? The basic notion is reasonably clear for scientific theories: it depends upon the entities said to exist by the quantificational structure of the theory. In the case of programs, we need to bring the specification into account.

> Let specifications A and B be given logically expressed specifications i.e., expressed in some logical language. Suppose that B is a logically stronger demand than A (logically stronger specifications imply weaker ones). If program P satisfies A but not B, whereas program Q satisfies B, then P is a *more parsimonious* solution to A than Q is.

This is one account of what parsimony means for programs that relies on an account of specification, and what it means for a program to meet a specification. When a program exactly matches a specification, it computes the same input-output relationship. So that minimality for programs is reduced to a notion of a minimal property enshrined in the specification.

16.3 Justification

Why is it rational to adopt simple designs? Principles of elegance and parsimony can be justified in epistemic or methodological terms:

- Why are simple programs and designs easier to grasp?
- Why does it make practical sense to adopt simple programs?

We have already commented upon the epistemic aspect of elegance: elegant programs are more perspicuous, are easier to grasp, and have more chance of being correct. Elegant programs can be more easily seen, or shown by argument, to satisfy their specifications; they are more explanatory in terms of how they solve the problem set out in the specification.

Parsimonious programs have an obvious methodological advantage: doing only what one is asked to do, and no more, reduces the amount of work. It also reduces the risk of errors. This is in line with Quine, who, in the case of theories, argues that parsimony carries with it pragmatic advantages, and that pragmatic considerations themselves provide rational grounds for discriminating between competing theories [191].

Chapter 17
MODULARITY

One of the methods advocated for achieving simplicity in design, for avoiding complexity, is modularization. Modular, structured, and object-oriented programming all aim at the construction of large programs and software systems by decomposition into smaller pieces [55]. Complexity is addressed by the separation of concerns: by decomposing a problem into smaller units, the complexity of individual units is reduced. Modularization is taken to increase the likelihood of correct designs: smaller units, when correct, are more transparently so. Presumably, as a result of breaking a problem into smaller units, the solutions to such units have a better chance of being elegant and parsimonious.

There are said to be other advantages to modular design. For one thing, modularization enables parallel work: by breaking the problem into parts, it allows work to be carried out by different agents. It also allows for future uncertainty: with a modular architecture, new module designs may be substituted for existing ones. It supports reuse, since modules that support some specified task may be employed elsewhere. From a methodological perspective, these seem to be clear advantages.

But what is modularity? What are its essential characteristics? Decomposing a problem into parts is just the beginning; the resulting units must have some desirable properties. For the computational notion, the literature enumerates a good number of these. Part of our task will be to explore these, and clarify their roles in any such characterization. In this regard, the following notions or characteristics are present in almost every exposition of modularity in the computational literature [25, 53, 81, 164]:

- encapsulation;
- independence;
- single function;
- cohesion.

© Springer-Verlag GmbH Germany, part of Springer Nature 2018

R. Turner, *Computational Artifacts*, https://doi.org/10.1007/978-3-662-55565-1_17

Methodological questions concern the mechanisms of modularization: How is it to take place? What techniques have been advocated for the process? Is there a best way to break a problem into components? What are the arguments marshaled to support the various methods of decomposition?

17.1 Encapsulation

Encapsulation is probably the most common attribute of modularization. In some form, it is to be found in the computer science notions of *module, abstract type, class, library* etc. However, as the following quotes demonstrate, there is some ambiguity about its interpretation:

> Encapsulation is a way of organizing data and methods into a structure by concealing the way the object is implemented, i.e. preventing access to data by any means other than those specified [65].

> In programming languages, encapsulation is used to refer to one of two related but distinct notions, and sometimes to the combination thereof: A language mechanism for restricting direct access to some of the object's components. A language construct that facilitates the bundling of data with the methods (or other functions) operating on that data [249].

Grady Booch, in his book on object-oriented design, emphasizes compartmentalization and hiding.

> Encapsulation is the process of compartmentalizing the elements of an abstraction that constitute its structure and behavior; encapsulation serves to separate the contractual interface of an abstraction and its implementation. Encapsulation is most often achieved through information hiding (not just data hiding), which is the process of hiding all the secrets of an object that do not contribute to its essential characteristics; typically, the structure of an object is hidden, as well as the implementation of its methods [25].

In their book on Design Patterns, Gamma et al. stress the hiding of information inside a unit or module:

> The result of hiding a representation and implementation in an object. The representation is not visible and cannot be accessed directly from outside the object. Operations are the only way to access and modify an object's representation [81].

In *Code Complete*, Steve McConnell insists that both bundling and hiding are central.

> Encapsulation (information hiding) is probably the strongest tool you have to make your program intellectually manageable and to minimize ripple effects of code changes. Anytime you see one class that knows more about another class than it should – including derived classes knowing too much about their parents – err on the side of stronger encapsulation rather than weaker [150].

These quotes are very representative of the ambiguity that surrounds the computational notion. Indeed, it seems clear that two different notions are involved:

- bundling;
- information hiding.

Bundling suggests that data and operations are packaged into a single unit that can be conceptualized as a single entity. In contrast, information hiding concerns access to the bundled information; not all the information in the bundle is equally accessible to all users. All the quotes mention these two aspects, and it would seem that both are taken to be part of the computational notion of encapsulation. While this may be so, the two are not the same thing. While some form of bundling seems necessary to facilitate any hiding, bundling information into a unit is possible without any hiding.

17.2 Bundling

Early instances of bundling include procedures where code is bundled together and given a name and a means of invoking it. This would seem to include procedural abstraction. But perhaps the most significant and recognizable notion is that of an abstract data type [146]. And this has its conceptual origins in the notion of a universal algebra. In actual programming languages, the various notions of a module take center stage [153, 214]. Indeed, contemporary languages contain many seemingly different representations of the basic idea: modules, classes, and libraries are common examples. But, in general, modules have two parts:

- signature;
- body.

A module interface or *signature* expresses the entities that are provided by the module, and the *body* of the module contains the *structure* or implementation. A module interface expresses the elements that are provided and required by the module, whereas the implementation contains the working code.

Figure 17.1 provides an example of a bundle based upon the notion of finite set. The signature declares the standard set-theoretic operations: membership, union, etc. The body contains their implementation as lists written in a functional language. There is something to check, i.e., that the module implementation satisfies its interface. In this regard notice that sets of type T are implemented as lists of type T. The operations thus follow suit. In particular, union takes two lists and returns a third that is the union of the input pair.

This is a simple illustration of bundling. The interface or functional specification is bundled together with its structure or implementation. Here implementation is not hidden; it is part and

empty : Set(T)
mem : Set(T)⇒ Set(T)⇒ Bool
add : Set(T)⇒Set(T)⇒Set(T)
rem : Set(T)⇒ Set(T)⇒Set(T)
size: Set(T)⇒ int
union: Set(T) ⇒Set(T) ⇒ Set(T)
inter: Set(T) ⇒ Set(T)⇒ Set(T)

---

empty = []
mem = List.mem
add x l = x :: l
rem x = List.filter ((<>) x)
rec size l =
match l with
| [] -> 0
| h :: t -> size t + (if mem h t then 0 else 1)
union l1 l2 = l1 @ l2
inter l1 l2 = List.filter (fun h -> mem h l2) l1

Fig. 17.1 Finite sets as a bundle

parcel of the bundle. In the language of technical artifacts, function and structure are lumped together in the module.

17.3 Information Hiding

Hiding structural details is common in everyday objects. For example, a dishwasher is to be used via its button interface; the user has no need to understand the machines inner mechanisms. If the machine was replaced by one with a different mechanism but the same interface, it could still be used as before. A facility of a programming or specification language that restricts access to some of the components of a datatype is the computational form of *information hiding*. It centers on the notion that a design decision should be hidden from the rest of the system. The computational concept appears to have been explicitly introduced into software design by Parnas [176].

> Every module in the decomposition is characterized by its knowledge of a design decision which it hides from all others. Its interface or definition was chosen to reveal as little as possible about its inner workings.

In our example module, the interface or signature is the first part of the module. We might facilitate hiding by separating the two aspects of the module into interface and implementation.

The central idea is that this is the only thing that is visible to other programmers. The actual implementation in terms of lists is hidden, not accessible for manipulation; it may be changed

empty : Set(T)
mem : Set(T)$\Rightarrow$ Set(T)$\Rightarrow$ Bool
add : Set(T)$\Rightarrow$Set(T)$\Rightarrow$Set(T)
rem : Set(T)$\Rightarrow$ Set(T)$\Rightarrow$Set(T)
size: Set(T)$\Rightarrow$ int
union: Set(T) $\Rightarrow$Set(T) $\Rightarrow$ Set(T)
inter: Set(T) $\Rightarrow$ Set(T)$\Rightarrow$ Set(T)

without affecting the function of the module. Information hiding gives the programmer the freedom to modify how the structure is fulfilled by a module without impacting upon its use.

A second example is to be found in the notion of a class. Consider again the class diagrams for a library system given in Fig. 6.1. Here the whole system has been divided into several classes. Each of the classes is a bundle. For instance, the *Book* module lumps together all the information about books into one class, and the *Patron* class does the same for users. Here the implementations of the operations are detached. The implementations are not formally part of the class specification but located somewhere else in the system.

Hiding is a core activity of software design that applies at every level of abstraction. It gives the engineer the freedom to modify how the implementation is given. This is especially valuable at points where the design (or even the requirements) is likely to change.

17.4 Independence

Information hiding is closely associated with another notion, namely *self-containment* or *independence*. Indeed, in cognitive science and the philosophy of mind, especially in the hands of Fodor [73], it is the central part of the notion of encapsulation. A notion of independence is to be found in the idea of modular programming which emphasizes separating the functionality of a program into independent, interchangeable modules. Independence allows change without affecting other parts of the system. Wherever the line is drawn between individual modules, the technique enables a separation of concerns, and allow programmers to develop components independently of each other: programmers may work on individual modules knowing only the functional specification of other modules. It is this notion of independence that supports the separate development of modules. But what is *independence*?

As part of his characterization of modularity [73], Fodor emphasizes the *character of information flow across computational mechanisms*. There are two aspects to this that are distinguished by the direction of flow. *Inaccessibility* concerns the internal inaccessibility of modules to central monitoring. Applied to the implementation component of a module, this is information hiding; the implementation is inaccessible or hidden. A module is *informationally encapsulated* if it cannot access information stored elsewhere in the containing system; it may only access the information contained in itself and in its inputs. But one must be careful about how to interpret this.

A weak version insists that modules do not have access to each other's implementations. A strong version would have it that modules cannot employ the operations declared in the interface of other modules. But this is too strong for computational purposes: modules do require access to each other's interfaces. In other words, one module can use another, but in doing so requires only its functional description; it does not require the details of its implementation. The extent to which modules depend on each other's interfaces is a measure of their mutual dependence or *coupling*.

17.5 Single Function

Often associated with encapsulation is a principle of simplicity: modules should be simple in the sense that they encode only one function. By tying modules to single functional demands, the tightness of match between problem and solution is optimized: minimality or parsimony is optimized. *The single responsibility principle* was introduced by DeMarco [53]. It insists that a class, module, or function should be responsible for a *single* aspect of functionality. Programs, modules, and classes, abstract types are more transparently parsimonious when they correspond to a single functional demand. For example, a program that sorts list of numbers has a single function. Each component should be responsible for a *single* aspect of functionality. Single responsibility also seems to be part of the biological notion of a module, where modularity refers to systems that consist of *discrete*, individual units.

However, it is not immediately clear what the computational notion amounts to. Some insight can be gleaned from the cognitive notion where a module that satisfies single functionality has a restricted subject matter, that is, the class of objects and properties that it processes information about is restricted:

> Domain specificity has to do with the range of questions for which a device provides answers (the range of inputs for which it computes analyses): the narrower the range of inputs a system can compute, the narrower the range of problems the system can solve–and the narrower the range of such problems, the more domain specific the device [73].

> Alternatively, the degree of a system's domain specificity can be understood as a function of the range of inputs that turn the system on, where the size of that range determines the informational reach of the system [34].

In a more computational arena, Bob Martin [155] also defines a responsibility as *a reason for change*, and concludes that a class or module should have one, and only one, reason for change:

In the context of the Single Responsibility Principle (SRP) we define a responsibility to be "a reason for change". If you can think of more than one motive for changing a class, then that class has more than one responsibility.

As an example, consider a module that constructs and prints a report. It seems to have two reasons to change. Specifically, the content or the form of the report could change. The single responsibility principle says that these two aspects of the problem are distinct. Consequently, they should be encapsulated in separate classes or modules. It would be a bad design that coupled two things that can change independently. A module focused on a single concern is more robust. The construction process and the printing process will not be independent if they belong to the same class or module.

But using the notion of *reason for change* appears to move us forward only to the extent that this notion is clearer than that of single functionality. For architectural design, a single aspect might be the design of a user interface or a database. But these might themselves be very complex components whose complexity only emerges at the next level of detail. A database that might be a module in a large software system has its own complex structure. Likewise, at the individual-program level, modules might form part of a library system. One module might be one that covers book lending to users, and another might facilitate their return. Within the overall system, these might be taken as single independent functions. But, at another level they might form part of a whole library system – the part that concerns the lending and returning of books. The internal complexity of the lending-returning module is hidden in the complexity of the whole library system. Indeed, it is part of the task of abstraction to hide the complexity in this way. The notion of *single function* seems to depend crucially on the level of abstraction. What seems to be a single function at one level of abstraction might have hidden complexity that becomes visible at the next level; what is a system at one level of abstraction might be seen as a module higher up.

Seemingly, the perception of modularity is closely related to the perception of the system. We cannot find the modular structure if we do not know the system to which the module belongs. Apparently, *single function* is a notion that must be determined or fixed relative to its containing system.

17.6 Cohesion and Coupling

Implicit in our discussion is the relation of dependence between modules. Modules interconnect in many ways. For instance, a module can call on data or operations from other modules or be called by other modules. *Coupling* is a measure of such interdependence. Good software design limits dependence to that which is somehow essential. Operationally, if one has to employ module B in order to use module A, then A and B are said to be *coupled*. This dependence inhibits change.

Cohesion is a measure of the connection between the operations and data within a single module; it is a measure of the internal coupling between the methods or the operations of a module, and reflects the strength of the relationships between its methods. High cohesion is associated with robustness, reliability, reusability, and transparency. Seemingly, high cohesion increases elegance.

What is referred to as *functional cohesion* occurs when the parts of a module are grouped together because they all contribute to a single well-defined task. Our *set module* has high functional cohesion in that it encapsulates the notion of finite sets. Where operations in a module are conceptually connected via a more embracing concept, as is the case with our finite-set module, they can be grasped in relation to the overall concept: our grasp of the individual operations is enhanced by its contribution to the overall meaning of the module.

Intuitively, low cohesion leads to modules that are difficult to maintain, test, reuse, or even understand. High cohesion often correlates with loose coupling, and vice versa [44]. The design problem is to produce a balance between coupling and cohesion. Intuitively, low internal dependence leads to higher dependence on other modules: individual modules that are too narrow become too tightly coupled to other modules. From this perspective, design is an optimization problem between lowering coupling and maximizing cohesion.

However, while the set example is clear, the general notion of functional cohesion is seemingly no more transparent than that of a single function. Again, what is taken as a cohesive module appears to be in the eye of the containing system.

17.7 Modularization

This brings us to the methodological question: how is the process of modularization to be carried out? What are the criteria to be used in dividing the system into modules? The above characteristics of modules give us the goals of modularization, together with some derived strategies such as maintaining a balance between coupling and cohesion. But much of the effectiveness of modularization depends critically on more pragmatic issues. There are two competing global modularization methods which might be characterized as follows:

- flow of control;
- identifying difficult design decisions.

The first divides complex tasks into simpler ones following the intended flow of control through the program. However, Parnas [176, 177, 178] argues that it is almost always poor design to begin the decomposition of a system into modules on this basis. In [177] he says:

it is almost always incorrect to begin the decomposition of a system into modules on the basis of a flowchart. We propose instead that one begins with a list of difficult design decisions or design decisions which are likely to change. Each module is then designed to hide such a decision from the others.

Instead of *flow of control* as the basis for modularization, he proposes that the designer should begin with a list of difficult or unstable design decisions. Each module is then designed to hide such a decision from the others. Such design decisions do not necessarily follow processing steps. For example, suppose we are designing a database with *input*, *search*, and *output* modules. All three modules depend on the internal representation used for the database. Consequently, each must be modified if this representation is changed.

Parnas's decomposition is driven by information-hiding concerns. In this case the database is something that is likely to change, and is common to many components. Hence, this information is hidden in a single database module that provides operations such as reading and adding data, and fetching and writing records. The rest of the program can now be written without knowing anything about how the database is implemented; to change the internal representation of the database, we need only substitute a different implementation of the database module:

```
read_record(file ,id ,record)
add_record(id ,record ,database)
get_record(id ,record ,database)
write_record(file ,id ,record)
```

Here we are at the edge of design theory rather than the philosophy of design. The philosophical task is to set the scene and clarify the issues, rather than make methodological decisions that are best left to the practitioners and theoreticians. We have tried to isolate the central concepts, clarify their meaning, and evaluate some of the associated arguments. Clearly, in this regard, there is more to do.

Chapter 18
FORMAL METHODS

Formal methods employ mathematical models to design software and hardware systems, and they employ mathematical proofs in an attempt to guarantee correctness. There are various phases to the formal approach:

1. A formal specification of the system design.
2. An investigation of the various formal properties of the constructed system.
3. Implementation of the various modules in the system.
4. Proofs that each of the modules satisfies its formal specification.

In the design phase, the engineer defines a system using a formal modeling language such as Z or VDM. This facilitates the second phase, in which she explores the model's mathematical properties such as the consistency of the formal model. The third stage is the coding or programming stage, and involves the implementation of the specifications of the various modules of the system. Finally, each of the modules must be proven to satisfy its specification.

This is the hard-nosed version. In reality few of the above stages are present in actual software construction. However, formal languages such as UML, and even the internal specification facilities of programming languages such as classes, might be employed throughout the construction. In this sense, all software development is formal to some degree. But the presence of formal proofs is the critical feature.

Several influential computer scientists have claimed that programming is, or should be, methodologically a mathematical activity. In various places, Hoare suggests that programs should, and can, be derived from their specifications in a mathematically precise way, employing program transformations based upon algebraic laws. He presents matters as follows:

> I hold the opinion that the construction of computer programs is a mathematical activity like the solution of differential equations, that programs can be derived from their specifications through mathematical insight, calculation, and proof, using algebraic laws as simple and elegant as those of elementary arithmetic [115].

© Springer-Verlag GmbH Germany, part of Springer Nature 2018

R. Turner, *Computational Artifacts*, https://doi.org/10.1007/978-3-662-55565-1_18

Programs are to be constructed via rules that preserve correctness. Of course, the process is not mechanical; it is a creative process just like the solution of differential equations. In *Mathematics of Programming* [115], Hoare advances several claims concerning the ideal practice of programming.

1. Computers are mathematical machines.
2. Programs are mathematical entities.
3. A programming language is a mathematical theory.
4. Programming is a mathematical activity.

The first and second we have addressed in Part II, and, in particular, in Chapter 5. The third claim, has been discussed in Chapter 8. The last claim underlies formal methods. The claimed advantage of such methods is that they offer the potential to deliver correct software; the claimed disadvantage is increased cost in time, money, and training. Are these two desiderata irreconcilable?

18.1 Formal Specification

We illustrate the issues with the logical approach to specification, based upon VDM. However, much the same core conceptual issues occur with any of the major specification languages.

Imagine the system being demanded is a suite of arithmetic functions for a simple calculator. Requirements analysis insists that we supply functions for addition, multiplication, division, greatest common divisor, and square root. We illustrate with the last two. The first stage involves specification: the following is a specification of $GCD(x : N, y : N, z : N)$ with true as the precondition and the following postcondition:

$$Divides(z, x) \land Divides(z, y) \land \forall w \bullet Divides(w, x) \land Divides(w, y) \rightarrow w \leq z$$

The postcondition asserts that z divides x and y, and it is the greatest number that does. The second example specifies the square root function. The postcondition asserts that y is the square root of x:

$$SQRT(x : Real, y : Real)$$
$$Pre : x \geq 0$$
$$Post : y * y = x \land y \geq 0$$

Although very simple examples, they suffice to illustrate the main features and problems associated with formal methods.

18.2 System Properties

Once the system has been specified, its formal properties need to be established. In a large system, each of the modules will need to be checked for consistency, coherence, and completeness. For example, we might wish to show that the square root specification defines a total operation, i.e., always gives a result:

$$\forall x : Real \bullet x \geq 0 \rightarrow \exists y \bullet SQRT(x, y).$$

Presumably, this must be established in the associated logic of the specification language. Each of the modules will generate such proof obligations – and do so before any coding has been done. This generates a fair amount of formal work; and work that is not to the taste, or indeed competence, of many programmers.

18.3 Programming

The next stage involves the construction of programs for each of the modules. The following formula embodies Newton's method to find the square root of a positive number n.

$$x_{k+1} = \frac{1}{2} \left(x_k + \frac{n}{x_k} \right)$$

At each stage, the new value is computed from the old one. This process continues until the old and new values (x_k and x_{k+1}) converge. The actual program generates better approximations to the square root until two successive approximations are within some given tolerance. This is the underlying algorithm that forms the basis of the following Pascal program [253]:

```
program squareroot
implicit none
real :: input, x, newx, tolerance
integer :: count
read(*,*) input, tolerance
count = 0
x = input
do
count = count + 1
newx = 0.5*(x + input/x)
if (abs(x - newx) < tolerance) exit
```

```
x = newx !
end do
end program squareroot
```

Next, consider, GCD. Here we employ the Euclidean algorithm. This is coded in Pascal as follows.

```
program gcd;
var a, b, answer : integer;
function gcd (a,b : integer) : integer;
var remainder: integer;
begin
while not (b=0) do
begin
remainder := a mod b;
a := b;
b := remainder;
end;
gcd :=a;
end;
```

We leave the reader to program the rest. In the direct formal approach, given these programs, the next task is to establish that they meet their specifications.

18.4 Module Correctness

Initially, we need to say what correctness amounts to. Consider the greatest common divisor. Via the semantics of its containing programming language, any program *gcd* for the GCD carves a function operating between the natural numbers, where, for simplicity, we shall assume that the types of the specification and programming languages are the same. In general, this may not be so. However, we can illustrate the main issues associated with correctness proofs with this simplification. The minimal correctness condition insists that:

$$(Divides(gcd(x,y),x) \land Divides(gcd(x,y),y) \land$$
$$\forall w.Divides(w,x) \land Divides(w,y)) \to w \le gcd(x,y).$$

The function or relation generated by the input/output behavior of the program must satisfy the constraints of the specification. How do we establish these assertions? In what follows, for the abstract specification

$$R(x : T, y : S)$$
$$Pre : \phi[x]$$
$$Post : \psi[x, y].$$

and the program P, we shall abbreviate the assertion

$$\forall x : T, y : S.\phi[x] \rightarrow (P(x) = y \rightarrow \psi[x, y])$$

as $x : T, y : S.\{\phi\}P\{\psi\}$.

This is a soundness condition for the program: given the precondition, the program satisfies the specification.

To make matters more concrete, we employ the Hoare system [114]. The rules of the system are based upon a basic judgment of the above form. Intuitively, the above asserts that if the predicate calculus assertion ϕ is true before the program P runs, then ψ will be true afterwards. The Hoare calculus supplies rules for all constructs in the language. We give the rules for the WHILE language. We shall leave the types of the variables implicit.

The axiom for assignment has the form

$$\{\phi[E/x]\}x := E\{\phi\}$$

where $\phi[E/x]$ is ϕ with E substituted for the variable x.

The rule for sequencing has two premises:

$$\frac{\{\phi\}P\{\psi\} \quad \{\psi\}Q\{\mu\}}{\{\phi\}P; Q\{\mu\}}.$$

So, for sequencing, the output state for the first program becomes the input state for the second. The rule for the conditional is given as follows:

$$\frac{\{B \wedge \phi\}P\{\psi\} \quad \{\neg B \wedge \phi\}Q\{\psi\}}{\{\phi\}\text{if } B \text{ then } P \text{ else } Q\{\psi\}}.$$

This states that a postcondition Q, common to the *then* and *else* parts of the program, is also a postcondition of the whole conditional.

In the *while* rule, ϕ is called the *invariant* of the loop. If the Boolean is true, and ϕ remains true when the program P runs, then ϕ is invariant through the whole *while* loop.

$$\frac{\{B \wedge \phi\}P\{\phi\}}{\{\phi\}\text{while } B \text{ do } P\{\neg B \wedge \phi\}}.$$

A proof involves stringing these together in order to demonstrate that the final state description agrees with the specification. This is a very brief overview of the general approach, but enough for our purposes. Below is a proof of correctness in the Hoare style for the GCD. Generally, such proofs are tedious and time-consuming to construct. And matters get worse as the complexity of the programs and software increases.

Let i_k and j_k be the values of i and j after k iterations. The crucial step in the proof is to find an *invariant* which describes the state of the program after each iteration. Choose the invariant S_k for each iteration as follows:

$$S_k = gcd(i_k, j_k) = gcd(a, b).$$

The proof now proceeds by numerical induction. For the base case we observe that before the loop, $i_0 = a$ and $j_0 = b$. Hence the invariant S_0 holds that is, $gcd(i_0, j_0) = gcd(a, b)$.

For the induction step, we assume that S_{k-1} holds. We are required to prove S_k, i.e.,

$$gcd(i_k, j_k) = gcd(i_{k-1}, j_{k-1}).$$

By the assignment rule, r is the remainder of the division of i_{k-1} by j_{k-1}, i.e.,

$$i_k = j_{k-1} \, and \, j_k = r.$$

So, by the definition of the division, we have :

$$i_{k-1} = Qj_{k-1} + j_k$$

for some integer Qj_{k-1}. There are now two cases. First, assume x divides both i_k and j_k. Since $j_{k-1} = i_k$, by the equation above, x also divides i_{k-1}.

Next, assume x divides both i_{k-1} and j_{k-1}. Then x divides i_k. Once again, using the equation above, x also divides j_k.

Therefore,

$$gcd(i_k, j_k) = gcd(i_{k-1}, j_{k-1}) = gcd(a, b)$$

as required.

By induction, the loop invariant still holds: $gcd(i, j) = gcd(a, b)$.

Some practitioners deny the practical feasibility of formal methods. While it may aid the construction of programs that meet their specifications, it adds another layer to the software construction process. And the formal approach also insists that properties of the specification be established before we proceed to this stage. So, in the fully fledged formal approach there are two stages that involves proof construction.

- investigation of the formal properties of the specifications;
- correctness proofs.

All in all, there is a lot of proving to do.

18.5 Transformational Programming

One way of attempting to alleviate matters is to employ program transformations. More explicitly, given an initial specification with preconditions and postconditions, the rules are applied backwards until the form of a correct program is uncovered. Transformations preserve meaning, and so correctness follows in their wake. However, programming still involves creativity through the application of the rules, but it is a disciplined creativity. This interpretation of the Hoare approach was further developed by Morgan [169], who provided a fully worked-out transformational approach for imperative-style languages. Progress towards an actual program occurs by refinement; we may, in logical terms, weaken the precondition or strengthen the postcondition. Specifications are eventually replaced by programs interpreted as relations between states.

This transformational approach is not restricted to the imperative paradigm: it has also been embraced in the functional style [109]. Here transformations do not necessarily involve a separate language of specifications. Instead, the initial functional program is somehow taken to be a specification of the problem. And while these transformations preserve correctness, they are primarily aimed at achieving more efficient programs, with the initial functional program taken as a specification of the problem. An approach that connects Z specifications and program transformations can be found in [110].

18.6 Constructive Programming

A radically different approach involves constructive logic. Martin-Löf [156] advocates constructive type theory as a means of deriving programs from (constructive) existence proofs. Constructive proofs contain constructions that are algorithms that witness the statement of the theorem. Turner [231] develops the Feferman approach to constructive mathematics, and uses realizability to abstract programs from constructive proofs. There has been considerable work on this approach over the last

10 years, with theorem-provers and interactive proof systems being employed for the construction of proofs.

18.7 Limitations of Formal Methods

Although they differ in technical details, all these methods put correctness as the major desideratum for program development. But the employment of such formal methods would seem to add much work to the actual process of software construction, and this is so whether one employs after-the-fact correctness techniques or transformational programming of some form. Despite the large amount of research and advocacy carried out by the formal methods community in the last 50 years, it appears that many practitioners view formal methods with skepticism.

Throughout the software development process there is a tension between the correctness and security of the overall system, and the speed and efficiency of constructing it. The process of constructing software via rigorous formal rules is seen by many to be unrealistic in practice. And this is taken to be so even with computer assistance.

This brings us to perhaps the most important and significant reason why formal methods have not provided a mathematical foundation for computer science practice. Formal systems are very difficult to use, especially for engineers untrained in the mathematical theories that underpin these formal systems. For instance, to employ an approach based upon set theory (e.g. Z) requires more than a little mathematical skill in the coding of systems in set theory. To the untrained, the mathematical nature of these languages makes it hard to model everyday systems. Formal methods introduce some form of computational model. As such the operations that may be employed are restricted to the model. For example, the Hoare system restricts one to those constructs that have mathematically clean rules of proof. However, these old systems are practically impoverished. Specifying systems is a technical endeavor: like the activity of programming, specification requires a rich and expressive means of representation. In particular, systems and languages of specification must possess a rich system of types and structuring mechanisms that enable them to support modularization and abstraction. Consequently, much research in formal methods is aimed at designing highly expressive specification languages. Unfortunately, this may lead to even more mathematical sophistication: the greater the mathematical sophistication of the framework, the greater the mathematical demands are made on the specifier. There is a tension between the skill set of the average software engineer and the mathematical sophistication of the required formal systems and tools. If formal methods become more common, engineers will have to learn type theory, modern algebra, and proof techniques. As Jim Horning [120] puts it:

> To treat programming scientifically, it must be possible to specify the required properties of programs precisely. Formality is certainly not an end in itself. The importance of formal specifications must ultimately rest in their

utility – in whether or not they are used to improve the quality of software or to reduce the cost of producing and maintaining software.

Is there a compromise? Are there are areas where formal methods are useful, and others where they accomplish little? Is it that formal methods are appropriate in some applications, such as safety-critical systems, but not in others? Is there an unbridgeable gulf between practice and the formal approach?

18.8 Programming as a Mathematical Activity

So, is programming a mathematical activity? Despite the overall skepticism of the practical community, even if they avoid formal methods, they cannot avoid reasoning: even if formal methods are not explicitly used, programming must involve some form of reasoning. In particular, Dijkstra [54, 59, 60] would have it that programming always involves some form of rigorous reasoning. For example:

> With respect to mathematics I believe, however, that most of us can heartily agree upon the following characteristics of most mathematical work: 1. Compared with other fields of intellectual activity, mathematical assertions tend to be unusually precise. 2. Mathematical assertions tend to be general in the sense that they are applicable to a large (often infinite) class of instances. 3. Mathematics embodies a discipline of reasoning allowing such assertions to be made with an unusually high confidence level. The mathematical method derives its power from the combination of all these characteristics; conversely, when an intellectual activity displays these three characteristics to a strong degree, I feel justified in calling it "an activity of mathematical nature", independent of the question whether its subject matter is familiar to most mathematicians [59].

This claims that the development of programs involves precise reasoning, presumably about the underlying type structure and control mechanisms of the programming language. There is little doubt that something like this is true. While programmers may not prove rigorous theorems about software, this does not mean that they do not reason when designing and testing programs; they must reason about abstract structures. Abstract types, polymorphic functions, classes, and objects are part of the modern programmer's toolkit. The programmer constructs programs with respect to their impact upon some abstract or virtual machine that captures the semantic content of the high level constructs of the language. It is this machine that supports reasoning about the types and operations of the language. At this most basic level, programming is a precise intellectual activity that involves precise reasoning about abstract, ultimately mathematical notions. This is a minimal claim that is hard to dispute; it insists that programming involves rigorous reasoning about some kind of formal objects. It does not demand that we employ any specific formal methods; it only insists that one cannot program without such abstract reasoning. The use of formal methods is a secondary consideration that aims to aid and supplement such reasoning processes.

It is important that these two claims are kept separate. One seems undeniable, while the other is controversial. While the use of formal methods might be taken to slow down the whole software process, what seems undeniable is that programming is an activity that involves precise formal concepts, and associated reasoning processes. Software development cannot be at odds with the latter; it is part of it. We should not identify mathematical activity with mathematics carried out inside a formal system. Almost no actual mathematics is. In this broader sense of reasoning about precise abstract concepts, programming is a mathematical activity. However, as we shall see in the last part of the book, complexity considerations complicate matters, and even challenge the notion of mathematical proof itself.

Chapter 19

THE DESIGN OF PROGRAMMING LANGUAGES

Good programming depends upon well-designed programming languages, and this has been recognized from the beginnings of the discipline. Indeed, the design of programming languages is one of the core activities of professional computer scientists. There are many properties that are taken to be necessary for a good design. Most critically, the language should aid the construction of well-designed and correct programs, and this leads to principles that more directly govern language design. However, we shall not try to provide a comprehensive overview and analysis of such design principles. Our objectives are more modest: we aim only to provide a commentary on three very general notions that many have taken to be the hallmarks of a good or successful design [168]:

- simplicity;
- expressive power;
- security.

Dealing with these should be enough to set the scene for more intensive studies on the philosophical aspects of language design. We shall deal with the third in a later chapter, where we consider the role of type inference in language design. Here we deal with the first two.

Taken at face value, simplicity would seem to entail fewer constructs, whereas expressive power seems to demand the opposite. The core design problem is to produce a language that is optimal for both. However, before we attempt to address this dilemma, we must investigate how these notions are defined and used when applied to programming languages.

19.1 Simplicity

As is the case with program and software design, many computer scientists place simplicity at the top of the list of characteristics of well-designed programming languages. MacLennan and Hoare express matters as follows:

© Springer-Verlag GmbH Germany, part of Springer Nature 2018
R. Turner, *Computational Artifacts*, https://doi.org/10.1007/978-3-662-55565-1_19

A language should be as simple as possible. There should be a minimum number of concepts, with simple rules for their combination [149].

Without simplicity, even the language designer himself cannot evaluate the consequences of his design decisions [116].

But what does "minimum" mean here? Are we talking about the smallest number of concepts that guarantees Turing completeness? Or is it a general aspiration or demand to keep the number of constructs to a smallish number? This seems to be the advice given by van Wijngaarden. In his paper on generalizing Algol, he writes:

In order that a language be powerful and elegant it should not contain many concepts. On the contrary by saying less one can say more, at least say more general things [247].

The latter brings generality into the picture. Apparently, by saying less, presumably keeping things simple, we may say more general things.

In our discussion of software design, we alluded to the distinction between two different senses of simplicity: syntactic simplicity and the ontological variety. There are variants of these notions that have been applied to programming languages. However, as we shall see, while there is overlap, they do not mean exactly the same thing for languages and programs.

19.2 Elegance via Uniformity

For programming languages, syntactic simplicity or elegance would seem to concern the syntax of the language. In crude terms, it refers to the number of basic constructs of the language (e.g. while loops, conditionals, guarded commands, and recursion). Accordingly, the simpler the language, the fewer its basic constructs. For example, an imperative programming language with procedures is less simple than a language without them. In other words, a form of Occam's razor is advocated as a design principle: if language L has fewer basic features than L^*, then it is rational to adopt the former. In this sense, the programming language C [130] is simpler than C++. But while this does capture some notion of simplicity, by itself it is too simplistic. In particular, it would appear to be directly at odds with expressive power: presumably, the more directly expressive the language, the more syntactic constructs will be required. A deeper notion of simplicity involves the *uniformity* of the constructs:

A well designed language should contain a relatively small set of primitive constructs that may be combined in a relatively small number of ways, where every possible combination of the language's constructs is legal. This is an orthogonality constraint that attempts to minimize exceptions to the rules. More generally, the language should be uniform in that similar features should look and behave in a similar fashion [149].

Languages that have a uniform structure are somehow simpler than those that have many special cases. More specifically, one way of reducing complexity is to make the actual mechanisms of combination, the ways of constructing new programs from old, *uniform*. Uniformity entails few, if any, special cases. This notion of simplicity can be illustrated with the case of the lambda calculus, the language that underpins functional languages such as Miranda and Haskell. This certainly has very few basic constructs: aside from variables it has none. And the ways of combining constructs are equally parsimonious: the facility to abstract on the variables (functional abstraction) and the facility to apply one thing to another (functional application):

$$t ::= x|\lambda x.t|tt.$$

These are easy to apply and there is a clear *uniformity* about their application. Indeed, there are no special cases or restrictions: abstraction and application may be uniformly applied to any expression in the language. Moreover, this language is Turing complete. So, despite its syntactic simplicity, it is as powerful as any programming language. However, Turing completeness is not the only measure of expressive power. Nor, in the present context at least, is it the most appropriate.

Indeed, this syntax is somewhat extreme in its uniformity. More often than not, for security purposes, actual languages are typed, and consequently impose more restrictions. It is here that security issues impact upon uniformity, and hence on simplicity. To illustrate this consider the typed lambda calculus. The types are either basic (B) or the type of functions from one type to a second.

$$T ::= B|T{\Rightarrow}T$$

Recall from our discussion of functional programming languages that the language is given by the following simple type inference rules:

$$\frac{d, x : T \vdash t : S}{d \vdash \lambda x.t : T \Rightarrow S} \qquad \frac{d \vdash t : T{\Rightarrow}S \quad d \vdash s : T}{d \vdash ts : S}$$

These rules are still uniform in application, and are restricted only by the type structure, i.e., leaving aside the type structure, application and abstraction apply uniformly throughout the language.

However, this restricts the language to the point where it is no longer Turing complete. And, while the latter is not a sufficient measure of expressive power, arguably it is a necessary one.

The optimization problem is to maintain security and uniformity without losing expressive power.

19.3 Parsimony

Ontological simplicity or parsimony concerns the underlying ontology of the language. It involves the number and generality of the concepts supported by the language. For Newtonian physics this concerns *mass* and *force*, and for Euclidean geometry it involves *points*, *lines* and *spaces*. This is Carnap's notion of the ontology of a theory: the things that the theory refers to or quantifies over. But how do we access the ontological commitments of a programming language?

Many languages are designed around a single ontological notion. For instance, functional languages are based upon the concept of a *function*. The lambda calculus is the pure version, while languages such Miranda and Haskell are examples of more employable ones. Object-oriented languages have focused on the notions of *class* and *object* (Eiffel), while the logic family (Prolog) employ the notion of a *logical proposition* or relation. In its own way, each is *ontologically parsimonious* in that its whole design is centered on one ontological notion: *functions*, *classes*, or *logical relations*. Of course, in practice, such purity is rare; most languages have a mixed-paradigm ontology. However, this only provides the top-level ontological commitments of a language, i.e., the fundamental notions that fix its paradigm. There are more delicate perspectives that drill down into the infrastructure of the language. More explicitly, the ontology of a language is determined by the ontological commitments of its semantic theory – by its underlying abstract machine and by its type system [187].

For example, the abstract machine for the WHILE language seems to commit us to some notion of abstract states, and operations on those states that allow us to change and examine the contents of its locations. This applies to both the operational and the denotational accounts.

Secondly, and more importantly for the representational capacities of the language, the type structure of the language commits us to some abstract notion of type. For example, the typed lambda calculus is committed to basic types, and the type of operations from one type to a second. In axiomatic terms, these types are fixed by the rules of membership and equality for their operators. In particular, for the lambda calculus they are fixed by the membership rules for abstraction and application, and their rules of equality or reduction. This would be the operational approach to ontology, where the axiomatic rules for the types fix their ontological nature [189]. Alternatively one might employ the denotational definition, in which case the ontology would be delineated by the semantic domains of the language [92, 97, 204]. But, in either case, it is the semantic definition of the language that records much of its ontological commitments.

So, what we take to be the semantic commitments of a language are dependent upon how we take the semantics to be given. If we think the semantics should be given in set theory, then the language seems committed to the fragment of set theory employed. In contrast, if we view types as fundamental (including if necessary some notion of state) and given by axioms and rules of membership and equality, then we are committed to the type theory of the language.

19.4 The Defense of Simplicity

It is said that simplicity aids the learnability and use of a language; it aids the construction and comprehension of programs. Presumably, the fewer and more uniform the constructs and types, the easier it is to learn: the more uniform the language, the fewer the special cases, and so there is less to learn. The motivation for this principle of uniformity is nicely defined by McConnell [150]:

> Regular rules, without exceptions, are easier to learn, use, describe, and implement.

Of course, this will be so if the uniform constructs are themselves clear and transparent. Some would argue that languages based on the lambda calculus are hard to learn because their underlying semantic notions are conceptually difficult. In other words, there is yet another notion of simplicity that pertains to the conceptual difficulty of the underlying semantic theory. For example, the set-theoretic semantics of the pure lambda calculus requires a set or domain that is isomorphic to some set of functions from the domain to itself [3]. To grasp what this amounts to, and to grasp its supporting mathematical notions, requires a level of mathematical competence that is not to be found in most programmers. The notion of ontological simplicity has an epistemological element that pertains to its semantic account. On the other hand, it might be argued that the complexity of the model is caused by enforcing a set-theoretic interpretation on an inherently operational notion. In contrast, the operational rules of the lambda calculus are easy to grasp and apply.

But there is another motivation for simplicity, and this involves the construction of programs. Uniformity is taken to aid elegant program construction. It is also taken to aid comprehension by others. Good design should result in a programming language that enables the human comprehension of programs.

Finally, simplicity should facilitate the construction of correct programs, i.e., the ability to minimize errors is one of the foremost design requirements. In this regard, Hoare [116] refers to a close association between the specification and the program. And this brings expressive power into focus.

19.5 Expressive Power

Most programming languages are Turing complete. Consequently, they are mutually intertranslatable. Every program in one language can be *correctly represented* in another. In particular, the lambda calculus is Turing complete. However, we lose this computational power in moving to the typed version. In particular, we lose the ability to represent recursion, which is available in the untyped language through the so-called Y operator,

$$Y = \lambda f.(\lambda x.f(xx))(\lambda x.f(xx)).$$

In the typed version, to recover matters, Y must be added as a new primitive. This is a simple illustration of how the more uniformity one has, the greater the expressive power – with fewer primitive constructs. This is one aspect of expressive power where a new primitive has to be added to make the language Turing complete.

But Turing completeness is not fine-tuned or sensitive enough for design. It is only one aspect of expressive power. A more relevant one is associated with ease of representation, the ability of a programming language to articulate a problem solution that is close to the original specification. We can get a better handle on things by posing the following question: when is one language more (representationally) expressive than another? Here there is some relevant computational literature.

Translation is not a sufficient condition for measuring expressive power; while it is a necessary condition, a more demanding notion is required. Felleisen [67] compares the expressive power of two languages as follows.

> Given two universal programming languages that only differ by a set of programming constructs, $\{c_1, \ldots, c_n\}$, the relation holds if the additional constructs make the larger language more expressive than the smaller one. Here "more expressive" means that the translation of a program with occurrences of one of the constructs c_i to the smaller language requires a global reorganization of the entire program.

To illustrate matters, consider *while* loops and *for* loops. Given conditionals and program sequencing, the second is definable in terms of the first in a direct way:

For i = 1 ,..., n do P = while i<n+1 do P

All other aspects being the same, if one language has both, and the other language has only *while*, the two are equally expressive. The new construct, the *for* loop, does not require a global reorganization of any program containing it. The addition of these new constructs is not only conservative, but also structurally so.

In contrast, a language with just *goto* statements would not be structurally conservative over one that had *for* loops: the translation of *for* loops into *goto* statements requires a complete reorganization of any program containing *for*. For example, the following.

For i = 1 ,..., n do i:=i+1

would need to be reorganized as follows:

i = 1;
L: P;
if i<n then do i:=i+1; Goto L else skip

There is also an ontological aspect of expressive power that is represented by the type structure of a language: a language that has a richer type structure is more able to directly model a wide range of problem domains. Consequently, it is more expressive than one with a restricted set of types. For example, a language with records as a data type can model databases better than one without. A language with modules can represent or package information that enables programs to be designed in modular ways. The demand for more types and more control constructs pulls in the opposite direction to syntactic simplicity. However, it is in this apparent conflict that the solution to good design resides.

In the next chapter, we shall look at three design principles that are aimed at resolving the conflict between simplicity and expressive power.

Chapter 20
SEMANTICS AND DESIGN

How do we design simple yet expressive languages? One piece of general advice about language design comes from one the founders of formal semantics, Christopher Strachey [219]:

> In these terms the urgent task in programming languages is to explore the field of semantic possibilities. When we have discovered the main outlines and the principal peaks we can set about devising a suitably neat and satisfactory notation for them, and this is the moment for syntactic questions. But first we must try to get a better understanding of the processes of computing and their description in programming languages. In computing we have what I believe to be a new field of mathematics which is at least as important as that opened up by the discovery (or should it be invention?) of calculus. We are still intellectually at the stage that calculus was at when it was called the "Method of Fluxions" and everyone was arguing about how big a differential was. We need to develop our insight into computing processes and to recognize and isolate the central concepts – things analogous to the concepts of continuity and convergence in analysis.

This suggests that we should employ semantic concepts in the process of design. Apparently, the field of semantic possibilities must be laid out prior to the design of any actual language, i.e., its syntax. More explicitly, the things that we may refer to and manipulate, and the processes we may call upon to control them, need to be settled before any actual syntax is defined. This is the semantics-first principle. According to it, one does not design a language, and then proceed to its semantic definition as a posthoc endeavor; semantics must guide design.

A more contemporary version of this advice comes in the form of three semantic principles [223] that are specifically aimed at designing simple yet expressive languages:

- correspondence;
- completeness;
- abstraction.

This chapter is given over to an analysis and evaluation of these principles that aim to optimize the design of simple yet expressive languages. But we are not suggesting that these are the only principles. For example, [149, 204, 205, 223] provide a zoo of such principles. Rather, we employ

© Springer-Verlag GmbH Germany, part of Springer Nature 2018
R. Turner, *Computational Artifacts*, https://doi.org/10.1007/978-3-662-55565-1_20

them as a means to expose some underlying conceptual issues pertaining to design, and to address the *simple yet expressive* requirement.

20.1 The Principle of Correspondence

The first principle is generated by the requirement of uniformity. In its original form, it is known as Landin's *principle of correspondence*. First, it demands that for each form of definition or declaration there exists a corresponding parameter mechanism, and vice versa. To illustrate matters, consider simple declarations of the following form:

Let $x=E$ in D.

Here E and D are expressions. Semantically this has the effect of substituting E for the identifier or variable x in D. According to the principle of correspondence, this must be matched by a parallel parameter-passing mechanism that takes the following form:

Define $G(x) = D$ in $G(E)$

Here $G(x) = D$ is a function definition. This has two parts. The function is defined in the following way:

Define $G(x) = D$

It may then be employed or called:

$G(E)$.

A programming language uses an evaluation strategy to determine when to evaluate values in declarations, and when to evaluate the arguments of a function call – and what kind of value to pass to the function. First, consider the declarations. Here x is assigned the value E in the expression D – which may contain x. Unfortunately, this is ambiguous: there are two evaluation regimes. One might evaluate E to obtain a value v. Then when D is evaluated, x is assigned the value v. This is an *eager* strategy. Using our operational-semantics notation, it may be formalized as follows.

$$\frac{<E, s >\downarrow v}{< \text{Let}\, x = E\, \text{in}\, D,\, s >\downarrow< D,\, s[v/x] >}.$$

Given a state s, we evaluate E to return a value v. Then D is evaluated in the state in which the variable x is assigned the value v. Alternatively, we might substitute E and evaluate D after the substitution has occurred–and x may not occur. This is a *lazy* strategy whose formalization takes the following form:

$$< \text{Let } x = E \text{ in } D, \ s > \downarrow < D[E/x], \ s > .$$

This makes a difference semantically. For example, if x does not occur inside D, we shall never evaluate E.

Consider the parallel function call. Here the function G has body D.

G(E).

The eager evaluation strategy is captured by the following rule.

$$\frac{< E, \ s > \downarrow v}{< G(E), \ s > \downarrow < D, \ s[v/x] >}.$$

Semantically, the value of $G(E)$ in a state s is the value of D in the state in which the variable x is assigned the value of E when evaluated in state s. Alternatively, we might substitute E and evaluate D after the substitution has occurred. This is a lazy strategy:

$$< G(E), \ s > \downarrow < D[E/x], \ s > .$$

The principle of correspondence demands that the two kinds of variable substitutions must agree: if declarations are evaluated one way (eagerly or lazily) then so should function calls and vice versa. The correspondence principle states that *the use of names should be uniform within a language.* Consequently, and this is the major motivation, the semantics of free identifiers/variables becomes independent of their mode of definition. Thus, the meaning of ... x ... is the same no matter what the context.

This is simplicity through uniformity. The central motivation for the principle is essentially the same as that which motivates the desire for compositional semantics. We intuitively assume a uniform semantics of the language; it reflects what we expect to happen because we implicitly adopt such a principle of correspondence. This is justified by the claim that a regular language with a uniform semantics is easier to learn and use.

Actual languages differ in satisfying these requirements. For example, Ada has nine different syntactic forms of declaration, each with a slightly different semantics. In contrast, the language Russell [23] has just a single form. In practice, the picture is very mixed. Placing constraints on the legal forms of parameters prevents the attachment of a uniform semantics to two obviously related mechanisms, parametrization and declaration.

20.2 Type Completeness

Type completeness places the type structure of a language at the core of the design process. A language is said to be *type complete* if and only if the following conditions hold:

1. Each operator of the language has a type, and the type of any composite expression is compositionally composed from the types of its components.
2. The type of an operator or expression may not depend on its surrounding context.
3. Every type is inhabited.
4. Any expression can be parametrized with respect to any free name having any type in the expression to yield an operation of an even more complex type.

The first condition implies that the type of a complex expression can be compositionally computed by the rules that govern the operators that form complex expressions from simple ones. This is also a demand for compositionality, and so a demand for ease of use and learnability. Condition 2 demands that the meaning of operators is uniform throughout the language. A simple example is provided by the type of identifiers; this may not depend on whether the identifier appears on the left or right side of an assignment statement. Condition 3 demands that there is an expression for every type of the language. Condition 4 insists that functions must be able to have parameters of any type, and to produce results of any type.

Probably the simplest example of a type complete language is the simple typed lambda calculus. Every closed term has a type. On the assumption that the variables are typed, every expression in the language has a type. Each lambda expression has a type. The type of composite expressions is composed from the types of the components (e.g., the operator and operand of an application) without regard for surrounding context. Assuming the base types have elements, every type is inhabited. For condition 4, we can always produce a more complex expression of higher type by lambda abstraction of any free identifier.

Type completeness guarantees that abstraction and application (function definition and calling) are uniformly applicable. Any name may function as a parameter which can be of any type – and an argument of that type can be constructed. Parametrization is a principle of abstraction. Type completeness allows us to define all forms of declarations in terms of the basic parameter/argument binding semantics, and ensures that any name that can be declared can also be a parameter. Consequently, by adopting the principle of correspondence, we can capture all the naming conventions of the language by a few basic notions, independently of what is being named.

The idea of using the type structure of a language as its framework suggests an approach to language design: instead of designing by adding new features, we should attempt to perform systematic enrichment of the underlying type structure, and to add new, general combining forms only when necessary. Type completeness brings uniformity to the syntax and semantics of the language, and eliminates the necessity for the introduction of mechanisms to handle special cases. So, once again, simplicity is obtained via uniformity. This is a *types-first* principle of language design that echoes the spirit of Strachey's *semantics-first* principle.

20.3 The Principle of Abstraction

A core aspect of expressive power concerns the ability of the language to support abstraction. Applied to languages, the principle of abstraction insists that all major syntactic categories should have abstractions defined over them. This is another principle of uniformity: we should be able to abstract over any semantically meaningful syntactic entity in the language. There are many variations of the basic statement, but the following by David A. Schmidt, is typical:

The phrases of any semantically meaningful syntactic class may be named [205].

The basic mechanism of control abstraction is a function or procedure. This amounts to lambda abstraction, the abstraction of the lambda calculus. In imperative languages, functions and procedures are abstractions of *expressions* and *commands,* respectively. Suppose that two chunks of code differ only in terms of a single expression: for instance, one may have 4 as a component and the other 6. Aside from this, the two pieces of code are identical:

Proc = body(4)
Proc' = body(6)

We may then, relative to that component, form an abstraction. This is called functional or procedural abstraction:

Proc(x)=body(x)

These forms of abstraction can be found in most contemporary languages. However, according to the above principle, there should also be abstractions over types (parametrized types). In particular, the following procedures are similar but differ in the types over which they operate.

Proc List(Num) = body(List(Num))
Proc List(Chr) = body(List(chr))

Here the abstraction is on types: by abstracting, we obtain a polymorphic procedure that requires types as arguments:

Proc List(X) = body(List(X))

This is the abstraction of the second-order lambda calculus [18]:

$$\lambda X.\lambda x.body(X, x).$$

Polymorphism allows programs to work for families of types. For example, the following type is the polymorphic type of lists:

$$\Pi X.List(X).$$

Languages are often designed to support some form of polymorphism – though not in the form of the full second-order lambda calculus.

Another instance of type abstraction is declaration abstraction. The value of a declaration abstract is called a *generic*, where in C++ generics are called templates. For example, we define a generic identity function by declaring the type of the parameters to be a parameter. In object oriented design [81], patterns are abstracted from the common structures that are found in software systems. Here, abstraction is the means of interfacing: it dissociates the implementation of an object from its specification; we abstract on the code itself. Abstraction applied universally is another principle of uniformity: it adds expressive power to the language but maintains simplicity through uniformity. Indeed, all three principles work together, and all are instances of simplicity and expressive power simultaneously acquired through uniformity.

20.4 Programming Languages as Mathematical Theories

We might use our semantic principles to design languages that are more likely candidates to be considered as mathematical theories. The whole enterprise of language design is a two-way street with theory and practice informing each other. In order to build pure computational theories, one must have some practice to reflect upon. Practice plus some theory leads to actual languages, which in turn generate new theories that feed back into language design. The various activities bootstrap each other. This finds the appropriate place for theory: it advocates a theory-first principle for each new generation of programming languages. This both endorses a more realistic interpretation of the semantics-first principle and increases the chances that the resulting theory will be mathematically kosher.

But there is a caveat. While there is a central role for mathematical elegance it is not an end in itself. Languages must aid programmers, and allow for their intellectual frailty. Mathematically sophisticated designers may too easily sacrifice human needs for mathematical elegance.

Chapter 21
DATA ABSTRACTION

Abstraction in all its forms has played a substantial role in the design of programming languages. It has motivated the various language paradigms, and has inspired, and is manifested in, the rich type structure and machinery of control to be found in contemporary languages. These include specification schemas, procedural abstraction, inheritance, iteration, recursion, polymorphism, abstract data types, modules, and classes. The quest for simpler yet richer means of abstraction is one of the central driving forces in language design. These language features both aid and constrain the style of programming and specification. In programming practice, abstraction mechanisms are employed to bring about simplicity in software design, where it is the programming language itself that largely supplies these mechanisms. Modularity divides the problem into parts; abstraction moves it to a less informationally crowded space. It provides the programmer and architect with a more amenable platform for problem solving and representation. For instance, a programmer involved in writing software for an accountancy package does not need to know how numbers are represented in the underlying hardware. In principle, she need only know the everyday rules that govern numbers, for instance, the simple rules for addition and multiplication. By hiding the layers that form the underlying implementation, the designer is better able to handle complexity. The programmer or architect works at the level of abstraction dictated by her working programming or specification language; and this is largely determined by the available representational and control mechanisms of the language. Complex implementation details below this level of abstraction are, in principle, ignored in the design process. This allows a separation of concerns, and facilitates interoperability and platform independence.

Abstraction in computer science, or what is called abstraction, is varied and various [90]. The term is used freely throughout the discipline, and covers a multitude of concepts. However, there are three notions that seem to be central:

- data abstraction;
- parametrization;

© Springer-Verlag GmbH Germany, part of Springer Nature 2018

R. Turner, *Computational Artifacts*, https://doi.org/10.1007/978-3-662-55565-1_21

- information hiding.

The first is located in the various mechanisms of abstraction that move us between the levels of representation located in contemporary computer science. The second, perhaps the oldest computational notion, is exemplified by so-called procedural and functional abstraction, generics and polymorphism. We considered this in the last chapter. Information hiding is also classified by many as an abstraction process, but we have already discussed this as a tool for program design in our discussion of encapsulation. Our objective in this chapter is to provide a commentary on the first notion. This is the one most commonly thought of as abstraction, and the one with a substantial philosophical pedigree.

21.1 The Traditional Account of Abstraction

Following Lewis [145], Burgess and Rosen [30] outline the three traditional ways of characterizing abstractness:

1. The way of abstraction;
2. The way of conflation;
3. The way of negation.

The second characterizes abstract entities as a distinction between particulars and universals. The third distinguishes between them by insisting that abstract entities have no spatial-temporal location. The first characterizes abstract entities as abstractions from more concrete ones. And since we are primarily concerned with the concept of abstraction in computer science, this seems the most relevant. This notion has a long history.

Aristotle formulated his conception of abstraction as an alternative to Plato's conception of mathematical objects. According to Aristotle, mathematical objects are created by the intellect by detaching or abstracting the form of individual or material objects. Mathematical objects (numbers, shapes, and geometrical solids) are created by this process. By abstraction, we bring into being objects that are general forms abstracted from individual and concrete things. However, they do not exist in their own right. This is to be seen in contrast to the Platonic notion, where mathematical objects exist independently of any such cognitive acts.

Later philosophers also viewed abstraction as an intellectual process, but one in which new *abstract ideas* are formed by reflecting upon several objects or ideas, and omitting the features that distinguish them. Locke, the originator of this notion, argued that general terms stand for *abstract ideas*, and that these are created through a process that separates these ideas from the spatial and temporal qualities of particular things. Locke [147] gives several examples to illustrate such a process. For instance, regarding the origins of the general representation of the color white, we are told:

> the same Colour being observed to day in Chalk or Snow, which the Mind yesterday received from Milk, it considers that Appearance alone, makes it representative of all of that kind; and having given it the name Whiteness, it by that sound signifies the same quality wheresoever to be imagin'd or met with; and thus Universals, whether Ideas or Terms, are made.

The representation for a simple quality is formed by ignoring some aspects or details. For instance, one is given a range of white things of varying shapes and sizes, and one ignores the respects in which they differ. In this way, we come to idea of *whiteness*. Something of this sort is what most people take to be the essence of abstraction: somehow some common properties form the focus of a new concept. A more contemporary version, which refers to *concepts* rather than ideas, is to be found in Skemp [207], where similarity recognition results in the embodiment of this commonality in a new concept:

> Abstracting is an activity by which we become aware of similarities among our experiences. Classifying means collecting together our experiences on the basis of these similarities. An abstraction is some kind of lasting change, the result of abstracting, which enables us to recognize new experiences as having the similarities of an already formed class. To distinguish between abstracting as an activity and abstraction as its end-product, we shall call the latter a concept.

Computer scientists actually echo some of these characterizations. One such is the account given by the computer scientist Tony Hoare [117]. He describes abstraction in the following way:

> Abstraction arises from recognition of similarities between certain objects, situations, or processes in the real world, and the decision to concentrate upon those similarities and to ignore for the time being the differences.

These quotes contain the essence of the common understanding of abstraction. In all accounts, using terminology as neutral as possible, there are two aspects to this traditional notion:

- a process of similarity recognition;
- the formation of a new idea or concept on the basis of these similarities.

Our aim is to provide an account of these two aspects as they are applied in the computational arena. But first we require a more logical account of them.

21.2 The Way of Abstraction

Crispin Wright and Bob Hale [101, 260, 259] have developed an account of abstraction that has its origins in an observation of Frege. Seemingly, many of the singular terms that appear to refer to abstract entities are formed by means of functional expressions: *the direction of a line, the number of books on the shelf.* While some singular terms denote ordinary concrete objects, such as the present king of France, the functional terms that pick out abstract entities are, on this account, characterized by an abstraction principle of the following form:

$f(a) = f(b)$ if and only if $R(a, b)$

Here R is an equivalence relation (reflexive, symmetric and transitive) and f is the functional term. For example,

The direction of a = the direction of b if and only if a is parallel to b.

The number of Fs = the number of Gs if and only if there are just as many Fs as Gs

According to Wright and Hale, abstraction principles have a special semantic status in that they appear to hold in virtue of the meaning of the expression on the right of the equation: they are implicit definitions of the terms on the left. It is claimed that in order to understand the term "direction" one has to know that "the direction of a" and "the direction of b" refer to the same entity if and only if the lines a and b are parallel. Moreover, the equivalence relation that appears on the right-hand side of the equation would appear to be *semantically prior* to the functional expression on the left. In particular, mastery of the concept of a direction presupposes mastery of the concept of parallelism, but not vice versa. Wright and Hale [101] put things as follows:

Standardly, an abstraction principle is formulated as a universally quantified biconditional – schematically:

$$(\forall a)(\forall b)(\Sigma(a) = \Sigma(b) \Leftrightarrow E(a, b))$$

where a and b are variables of a given type (typically first–or second–order), "Σ" is a term forming operator, denoting a function from items of the given type to objects in the range of the first-order variables, and E is an equivalence relation over items of the given type. What is crucial from the abstractionist point of view is an epistemological perspective which sees these principles as, in effect, stipulative implicit definitions of the Σ-operator and thereby of the new kind of term formed by means of it, and of a corresponding sortal concept. For this purpose it is assumed that the equivalence relation, E, is already understood and that the kind of entities that constitute its range are familiar – that each relevant instance of the right hand side of the abstraction, E(a,b), has truth-conditions which are grasped and which in a suitably wide range of cases can be known to be satisfied or not in ways that, for the purposes of the Benacerrafian concern, count as unproblematic. In sum: the abstraction principle explains the truth conditions of Σ-identities as coincident with those of a kind of statement we already understand and know how to know.

This provides an account of the similarity recognition stage. The second part, the concept formation stage, is covered by the following. When "f" is a functional expression governed by an abstraction principle, there will be a corresponding kind K_f such that

x is a K_f if and only if, for some y, $x = f(y)$.

The availability of abstraction principles meeting these conditions may be exploited to yield an account of the distinction between abstract and concrete objects. The simplest version is the following:

x is an abstract object if (and only if) x is an instance of some kind K_f whose associated functional expression "f" is governed by a suitable abstraction principle.

21.3 Russell's Paradox

Unfortunately, the strong version of this account, which purports to identify a necessary condition for abstractness, is problematic. Pure sets are paradigmatic abstract objects. But it is not clear that they satisfy the proposed criterion. According to informal set theory, the functional expression "set of" is indeed characterized by a putative abstraction principle.

The set of Fs = the set of Gs if and only if, for all x, (x is F if and only if x is G).

But instances of this principle lead to Russell's paradox. In the above case, the class of predicates permitted on the right-hand side is left open, and so potentially may employ the very notion of a set itself. Hence Russell's concept of a set that is not a member of itself leads to a paradox since it is, by the principle of abstraction, a set that both is and is not a member of itself. In contemporary mathematics, the concept of a set is introduced axiomatically rather than by an abstraction principle, and according to [29] it appears unlikely that the mathematical concept of a set can be characterized in the latter way. We shall say more about the kinds of abstraction employed in computer science, and how they might avoid the kind of impredicativity that underlies Russell's paradox.

21.4 Data Abstraction

In computer science, new levels of abstraction are determined by the creation of new computational platforms. These are built into the languages themselves, and their discovery forms a major objective of programming language design. The notion of an abstract data type was first proposed by Liskov and Zilles in 1974 [146] as part of the development of the CLU language. One of the characteristics of this form of abstraction involves the suppression of information. Some data types contain less implementation information than others: they are at a greater distance from the underlying physical machine.

The abstraction step from finite lists of some type T with a strict linear ordering (e.g., Numbers) to finite sets illustrates this kind of abstraction process. The *computational information* in sets is less than that held in lists: the elements of a list come in a given order, and this ordering forces the programmer to be concerned with how elements are to be processed. Operations such as sorting a list of numbers into numerical order are only meaningful in the presence of such ordering. The latter is only important at a certain level of detail where the programmer is concerned with sorting algorithms and their efficiency. However, for set processing, these details are not visible; sets do not support, nor do they give rise to such processing. As it is put by Gutting [99],

> The essence of abstractions is preserving information that is relevant in a given context, and forgetting information that is irrelevant in that context.

The move from lists to sets is a paradigm case of data abstraction. But what form does the abstraction take? If we ignore the order of elements in a list, then we obtain a collection that is characterized by the elements alone: two lists are identified when they have the same elements. This is the common feature, and it is expressed via the notion of list membership:

> *Two lists are **similar** if they have the same elements.*

Two lists are similar exactly when they have the same elements in any order, and regardless of how many copies each contains. This is the *axis of abstraction*, the means by which we identify similarities. But how does this enable the *abstraction* of the notion of a finite set? If we apply the way of abstraction, we obtain the following:

> The set of a = the set of b if and only if a has exactly the same elements as b.

More formally, for $a : List(T)$, $b : List(T)$ we have,

$$set(a) = set(b) \text{ iff } \forall x : T.member(x, a) \leftrightarrow member(x, b)$$

where *member* is the membership relation for lists. According to the way of abstraction, the set of elements in a list has the status of a new abstract notion. This captures the computational notion of a finite set:

x is a finite set of type T if and only if, for some list y, $x = set(y)$.

These principles may be used not just to abstract the underlying set-theoretic structure, but to define the operations that go with it. In particular, we write

$$a \epsilon set(b) \triangleq member(a, b)$$

We may also define the basic operations on finite sets that form the set–theoretic structure of an abstract data type. For instance, we have the following notions:

$$a \bigcup set(b) \triangleq set(a * b)$$
$$\varnothing \triangleq set([])$$
$$set(a) \subseteq set(b) \triangleq \forall x.member(x, a) \rightarrow member(x, b)$$

where * concatenates the lists, the empty set is identified with the empty list etc. As long as our equivalence relation is a congruence with respect to these new defined operators, the theory is coherent.

The move from lists to sets is motivated by the desire to move to a new abstraction level where complexity is reduced. Lists are more complex because of the ordering. But they are also prior in the sense that they are closer to the machine; closer to the machinery supplied by the computer itself. Moreover, we know how to implement them.

Data abstraction facilitates the introduction of new type constructors. In computational terms, we begin with some collection of types that are closed under a type constructor $List(T)$, the elements of which are lists of objects of type T. This rule is restricted to those types T that come equipped with a strict linear ordering. We then apply the way of abstraction. In type-theoretic terms, via abstraction, we add a new type constructor.

$$Set(T)$$

whose members are finite sets of type T. This is then made available at the next level of abstraction, i.e., this new type constructor may be employed in the new class of types. In this way impredicativity is avoided in the formation of new types by abstraction. Abstraction levels are built layer by layer to avoid the impredicativity underlying the Russell paradox.

21.5 Principles of Abstraction

More generally, in the process of abstraction, we are essentially forming new abstract data types. We begin with some given type T together with an equivalence relation R on T. We then employ the way of abstraction to form a new type:

$$Abstract(T, R)$$

This is governed by the rules:

$$\frac{t : T}{abstract(T) : Abstract(T, R)} \qquad \frac{abstract(T) : Abstract(T, R)}{t : T}$$

$$\forall x : T. \forall y : T. (abstract(x) = abstract(y)) \Leftrightarrow R(x, y) \, .$$

This approach fits the process of abstraction in computer science, where new abstract types generate new layers of abstraction.

It also mirrors the dynamic approach to abstraction [220] in which abstraction results in new objects. In the dynamic approach, every question about the objects obtained by abstraction reduce to questions about the old objects. In our case, this is made explicit by a translation (*) between the old and new theories which is induced by the following clauses.

$$(abstract(x) = abstract(y))^* = R(x, y)$$

$$abstract(x)^* = x$$

The new theory is compiled into the old one. So, the type-theoretic way of abstraction provides a way of implementing the new abstract notions in terms of the old ones.

21.6 Implementation as a Basis for Abstraction

Although we have adopted the general schema of abstraction advocated by Wright and Hale, we have not explicitly adopted the perspective that the right-hand side of the equation is *semantically prior* to the functional expression on the left. Instead we have opted for a more concrete account of computational abstraction. The right-hand side must, in some sense, contain more information. For example, the list concept is informationally richer than that of sets. But there is a more concrete way of unpacking matters that refers to the underlying machine.

In implementation terms, we must already know how to implement the concepts from which the new notion is abstracted. In particular, lists may be implemented as linked lists which can be represented directly in the store of a concrete machine. Each memory cell contains two pieces of information: the first is the head of the list, and the second is a reference or pointer to the tail of the list, which may be stored somewhere else in memory. We may then impose an equivalence relation on these chunks of memory; two such linked lists are equivalent if they contains the same elements in the same order. This can be imposed by insisting that the first of the two linked lists contain the same element, and then insisting (recursively) that the rest are equivalent – which will unpack by checking that the first of these two contains the same first element, and so on. The recursion is coherent since all the lists are finite, and so the process of checking similarity will always terminate.

$$l_1 \overset{\circ}{=} l_2 \Leftrightarrow (first(l_1) = first(l_2) \wedge contents(second(l_1), s) \overset{\circ}{=} contents(second(l_2), s)$$

where $contents(second(l), s)$ returns the contents (the linked list) of the cell whose location is $second(l)$ in the state s. Now we define, by way of abstraction, the following.

$$List(l_1) = List(l_2) \text{ iff } l_1 \overset{\circ}{=} l_2$$

This provides the notion of a list abstracted from the notion of linked lists. Again the notion of linked lists is prior, in the sense that linked lists are rooted in a given physical implementation. This provides the platform for a new level of abstraction, a new abstract platform on which to base the next abstraction – from lists to sets.

Computational abstraction analyzed in this way sees the right-hand side of the Fregean abstraction principle as being prior in the sense of *known implementability*. Our account is essentially epistemic. The crucial property of the right-hand side is that it is known how its terms may be implemented.

Each layer of abstraction moves us farther from the machine: from linked lists, to lists, to sets. Since we know how to implement lists as linked lists, we can implement sets as lists. Hence, we know how to implement sets.

21.7 A Theory of Computational Abstraction

This is only a sketch of a theory of abstraction for computer science. However, the outlines of the theory are relatively clear.

- Levels of abstraction are generated by the way of abstraction.
- The way of abstraction generates new type theories from old ones.
- The right hand side of the abstraction principle is epistemically prior in the sense that it is an implementable type.
- The new theory is implementable in the old one.

A more complete development of this theory will appear elsewhere [238].

Part V
EPISTEMOLOGY

It has been claimed that one of the most significant contributions to epistemology is the mathematical analysis of computation. For example, the following is Gödel's observation of 1946:

> Tarski has stressed in his lecture (and I think justly) the great importance of the concept of general recursiveness (or Turing's computability). It seems to me that this importance is largely due to the fact that with this concept one has for the first time succeeded in giving an absolute notion to an interesting epistemological notion, i.e., one not depending on the formalism chosen [50].

Does the analysis of computation provide a significant contribution to epistemology? Why is Turing's analysis so significant?

What is it to know a programming language? Presumably, it involves both *knowing that* and *knowing how* knowledge. Knowing how to construct elegant programs and software is an instance of the latter, whereas knowledge of its syntax and semantics is an instance of the former.

What kind of knowledge do computer scientists have when they have established the correctness of their programs? Do they have mathematical knowledge? Are there different notions of correctness operating that return different kinds or forms of knowledge?

These are the concerns of this final part of the book. Inevitably, such questions will also give rise to semantic, methodological, and ontological issues.

Chapter 22
COMPUTABILITY

The work of Turing and Church has had a profound influence on the foundations and development of computer science. Church's work on the lambda calculus not only provided one of the central mathematical accounts of computation, but also paved the way for the development of the functional paradigm, and the semantics of programming languages. Turing is regarded as providing the definitive account of computation, inspiring the design of early computers, and giving birth to artificial intelligence.

In this chapter we concentrate on their common contribution, one of the great foundational pillars of theoretical computer science, indeed of computer science itself, namely the mathematical analysis of computation.

22.1 Mathematical Modeling

The original motivation for a mathematical analysis of computation came from mathematical logic. Its origins concern the decidability of the first-order predicate calculus: could there be an algorithm, a procedure, for deciding the truth or falsity of an arbitrary sentence of the logic (the *Entscheidungsproblem*). For first-order logic, this question was posed by David Hilbert and Wilhelm Ackermann in the twentieth century. Given the completeness of first-order logic, the problem is equivalent to deciding its provability. In order to address this question, a rigorous model of the informal concept of an effective or mechanical method in logic and mathematics was required.

Any mathematical or formal account must satisfy two constraints:

1. Every formally computable function is informally computable.
2. Every informally computable function is formally computable.

The first corresponds to informal soundness, and the second is informal completeness. These are demanding criteria, and mathematical modeling of this kind is parallel to the formalization of

© Springer-Verlag GmbH Germany, part of Springer Nature 2018

R. Turner, *Computational Artifacts*, https://doi.org/10.1007/978-3-662-55565-1_22

mathematical inference i.e., first-order logic is intended as a formalization of ordinary mathematical inference.

Alonzo Church and Alan Turing published independent papers that purported to demonstrate a general solution to the *Entscheidungsproblem*. Indeed, a good number of solutions were proposed that all turned out to be extensionally equivalent. Our objective is to place these accounts within the setting of contemporary computer science. From this perspective, there are two approaches to the formalization of computability:

- the programming language approach;
- the machine approach.

22.2 The Programming Language Approach

One early notion of formal computability, enshrined in the General Recursive Functions, was provided by Gödel and Herbrand in the spring of 1934. Church and Kleene also formulated the lambda calculus in the early 1930s and Kleene went on to demonstrate that these two notions were equivalent. This gave rise to two extensionally equivalent versions of the Church computability thesis:

> The computable functions are precisely those computable by lambda terms/general recursive functions.

From a computer science perspective, the general recursive functions constitute a rather unsophisticated functional programming language. There are a few basic numerical functions, such as successor, and complex ones generated by *composition, primitive recursion*, and a form of functional definition, (*minimization*), one that is naturally implemented as iteration.

Arithmetic texts contain computations involving addition, multiplication, and exponentiation. These involve special cases of composition and primitive recursion. As such, they reflect aspects of ordinary mathematical practice. However, the general recursive functions are a very big class, the majority of which have never seen the light of day. Even the small subset called the *primitive recursive functions* go way beyond ordinary arithmetic as it is practiced in the high street. However, we can see how to compute by hand with all of the strategies of the general recursive functions: we can see how to follow the rules with pencil and paper. Soundness would seem not to be problematic. However, completeness is quite unclear – or it would have been in the 1930s.

The lambda calculus may be seen as an even more primitive functional language. Indeed, in so far as functional languages are implemented in the lambda calculus, the latter is the machine language of functional programming languages. How would one demonstrate that it satisfies constraints 1 and 2? For constraint 1, again there is a plausible argument that every lambda computation is informally computable. Why? Because we can actually perform lambda reduction by hand. However, completeness seems even harder to demonstrate than for the general recursive functions. In particular, to show that every informally computable function over the natural numbers is lambda computable, one needs to code the numbers in the formalism of the calculus. One standard representation is the following, where f is applied n times:

$$n = \lambda f.\lambda x.f^n(x)$$

With this representation, addition is then represented as follows:

$$\lambda f.\lambda g.f(\lambda wyx.y(wyx))g$$

So, addition is easily shown to be representable. But how on earth do we demonstrate this for all informally computable functions? Completeness would seem to be problematic in both cases. However, historically, a further step was taken.

22.3 Invariance

Kleene proved the equivalence between the lambda computable functions and the general recursive ones. In computer science terms, such proofs involve the construction of a compiler or interpreter between the two formalisms, together with a proof of their correctness. This offers indirect justification for the informal computability of lambda terms and of the general recursive functions. This goes some way towards demonstrating that the characterized notion of computability is independent of any particular formalism. This is normally referred to as the *invariance condition.*

Of course, such proofs only offer absolutely compelling evidence on the assumption that one of the target formalisms is a sound and complete formalization of the intuitive notion. Indeed, despite the equivalence between the general recursive functions and lambda definability, according to Church, Gödel himself was not convinced that either of these two accounts gave a satisfactory analysis of the notion of *computable* or *mechanical*. Church, in a letter to Kleene, wrote:

> In regard to Gödel and the notions of recursiveness and effective calculability, the history is the following. In discussion with him [*sic*] the notion of lambda-definability, it developed that there was no good definition of effective calculability. My proposal that lambda-definability be taken as a definition of it he regarded as thoroughly unsatisfactory [131].

What Gödel eventually took to be a satisfactory account was provided by Turing.

22.4 The Machine Approach

Instead of defining computation in terms an informal programming language for mathematical practice, Turing aimed at modeling the limits of human computation [228]. Turing imposes a small number of constraints on the process of informal computation, spelled out in terms of what a human can do with pencil and paper. The heart of these constraints concerns the finiteness of the resources available to such a *human computer*. For example, at any given time, she has a finite sheet of paper to compute on. It is these informal boundaries that are meant to map the limits of informal computation, and thereby engulf completeness.

With these informal boundaries laid out, Turing went on to construct a formal model of them: now called Turing machines. The accepted consensus is that his account is the definitive one, and this is summarized in the Turing thesis, one formulation of which is the following.

- Turing machines can do anything that could be described as rule of thumb or purely mechanical.

Turing proved that his machines and the lambda calculus are extensionally equivalent. So invariance remains intact. Even today, every program written in an existing implemented programming language is Turing computable, and, conversely, all so-called general-purpose programming languages are Turing complete, i.e., they contain all the control constructs necessary to simulate a universal Turing machine. Indeed, the requirements for completeness are minimal. For example, the WHILE language is Turing complete – it contains sequencing, conditionals and iteration.

Turing provided an intensional characterization of the notion of computability: one that set limits on the notion of a human mechanical procedure or algorithm.

22.5 Semantics and Turing Machines

An important psychological part of the appeal of Turing's machines concerns the nature of their basic instructions; these are said to be *atomic* in the sense that they can be performed *without thought*. Turing puts matters as follows.

> Instructions given the computer must be complete and explicit, and they must enable it to proceed step by step without requiring that it comprehend the result of any part of the operations it performs. Such a program of instructions is an algorithm. It can demand any finite number of mechanical manipulations of numbers, but it cannot ask for judgments about their meaning. An algorithm is a set of rules or directions for getting a specific output from a specific input. The distinguishing feature of an algorithm is that all vagueness must be eliminated; the rules must describe operations that are so simple and well defined they can be executed by a machine [229].

The instructions must be complete and explicit, and the human computer is not required to *comprehend* any part of the basic operations; she cannot ask for judgments about their meaning. Other commentators seem to claim much the same:

> The intuitive notion of an algorithm is rather vague. For example, what is a rule? We would like the rules to be mechanically interpretable, i.e. such that a machine can understand the rule (instruction) and carry it out. In other words, we need to specify a language for describing algorithms which is general enough to describe all mechanical procedures and yet simple enough to be interpreted by a machine [243].

Is it that these atomic instructions have a meaning but may be performed without understanding? For the sake of argument, assume that they are not meaningless; they have semantic content. This does not mean that eventually the human computer will be able to perform atomic instructions without thought. The human computer may be able to. But, to begin with at least, the meaning must have been deployed. How else could one carry them out? Turing cannot be arguing that the human computer can be trained to do this without thought. This would make little sense, since the atomic operations are the most primitive. If the computer is to be trained to learn these, then there must be more primitive ones that form the basis of the learning process. And the atomic instructions are the most primitive ones. So it is hard to make out how, if they have meaning, how the human computer can initially proceed without thought. So, if the atomic instructions are taken to have meaning, it is hard to see what can be meant by the claim that the human computer does not comprehend the result of any part of the operations it performs.

Seemingly, we must assume that Turing's atomic instructions are *meaningless*. However, Shanker has argued, in terms of rule-following considerations, that this cannot be so. If Turing machines are collections of rules, which they surely are, the instructions cannot be meaningless: rules cannot be meaningless.

> But given that this is a genuine rule it is anything but meaningless. Granted, it is so simple that one can envisage an agent doing it mechanically after a short while, and that is the problem that will be looked at last. For the moment we need only see that, however simple this rule might be, it does indeed tell one what to do [202].

Indeed, from a computer science perspective, there is a devastating consequence of any such meaninglessness thesis: if all atomic instructions are meaningless, then there is a tension with the compositionality constraint on semantic theory. On the assumption that all basic or atomic instructions are meaningless, have no semantic content, it is hard to maintain any compositional theory of meaning. Via compositionality, a Turing machine program is a collection of conditional expressions that inherits its semantic content from its atomic instructions. And so, if all the atomic instructions are meaningless, the compositionality thesis would lead us to conclude that all programs in the language of Turing machines are meaningless, i.e., the collection of instructions that constitute any Turing machine, is meaningless. But this is absurd.

So it would seem that the operations are not meaningless, yet they can be immediately grasped without thought. It is hard to see how both can be true.

22.6 Turing Machines as Technical Artifacts

Why did Turing insist on this *meaningless* requirement? Presumably, because he takes such rules or programs to be mechanically implementable. Shanker puts this in terms of the removal of normative content:

> The only way to remove its normative content is to treat it as a description of the causal events that occur in the mind/program of the computer which trigger off certain reactions. Thus Turing introduced the premise that "the behaviour of the computer at any moment is determined by the symbols which he is observing, and his state of mind at that moment". On this picture the computer's "state of mind" is the causal intermediary between "observed symbols" and subsequent action [202].

Shanker is suggesting that a Turing machine, an abstract machine, is to be taken as a description of a (human) computer. This seems correct. But then Turing machines have no normative content.

However, there is a way of bringing the normative role of atomic operations back into the picture. Once the machine has been described, an intentional shift occurs. Atomic instructions have their obvious meaning, and are intended to be normative in the sense that they are to be taken as functional specifications of part of a physical computer. Abstract Turing machines are to provide the functional interpretation of Turing's physical computing machine. And they are able to play this role because, as abstract machines, they have perfectly sound semantic content. The role of atomicity is then clear: if we are to build such a device, a physical simulation of a human computer, part of the engineering task is to ensure that the atomic instructions are mechanically implementable, and this is part of the design requirements.

22.7 Machines and Rules

Do physical machines compute? Indeed, do they program or prove theorems? Do they follow rules? In the long discussion on rule following in [255], Wittgenstein continually reverts to the example of calculation in order to ask the question: Under what circumstances can I say of a physical machine that it computes?

> Does a calculating machine calculate? Imagine that a calculating machine had come into existence by accident; now someone accidentally presses its knobs (or an animal walks over it) and it calculates the product 25×20 Imagine that calculating machines occurred in nature, but that people could not pierce their cases. And now suppose that these people use these appliances, say as we use calculation, though of that they know nothing.

Thus they make predictions with the aid of calculating machines, but for them manipulating these queer objects is experimenting. These people lack concepts which we have; but what takes their place? Think of the mechanism whose movement we saw as a geometrical (kinematic) proof: clearly it would not normally be said of someone turning the wheel that he was proving something. Isn't it the same with someone who makes and changes arrangements of signs as [an experiment]; even when what he produces could be seen as a proof?

There are no causal connections in a calculation, only the connections of the pattern. And it makes no difference to this that we work over the proof in order to accept it. That we are therefore tempted to say that it arose as the result of a psychological experiment. For the psychical course of events is not psychologically investigated when we calculate. You aren't calculating if, when you get now this, now that result, and cannot find a mistake, you accept this and say: this simply shows that certain circumstances which are still unknown have an influence on the result. This might be expressed: if calculation reveals a causal connection to you, then you are not calculating. What I am saying comes to this, that mathematics is normative [256].

On this view, calculation and computing are mathematical activities. In contrast, what we do when we run programs on physical machines is experimentation. Wittgenstein draws attention to how the essential normativity of mathematics underpins the notion of calculation.

Computations and programs, as physical devices that run on actual machines, are computational artifacts. The activity of constructing a symbolic program from its specification is a rule governed mathematical activity. Whether machines can be said to carry out such an activity, whether they can be said to compute or program, whether they can be said to prove theorems, depends upon their ability to act as intentional agents, and not just on their ability to get the calculations right. This claim will form the focus of our discussion of the epistemological aspects of computer science.

Chapter 23
FEASIBLE COMPUTATIONS

Computational artifacts must perform their function. Part of this involves constraints upon their practicality: programs should involve computations that can be carried out in a *reasonable* time using *reasonable* resources. But what is *reasonable*? Presumably, a program that on average takes more than the age of the universe to return its results is not reasonable.

23.1 Computational Complexity Theory

The standard way of measuring matters is in terms of Turing machines, where the basic definitions of time and space complexity were first systematically formulated in 1965 by Hartmanis and Stearns [107] in a paper called "on the computational complexity of algorithms". Formally, a function f is a measure of the time complexity of an algorithm if, given an n-bit instance of the problem as input, it can produce a solution in time $f(n)$. For example, there are a considerable number of known algorithms for sorting lists, and they vary in complexity. Bucket sort is $O(n)$, that is, it is a some linear function of the input. Bubble sort, insertion sort and selection sort are $O(n^2)$ – some quadratic function of the input. Mergesort is order $O(n\log n)$, etc. Arguably, from a practical standpoint, this is the most relevant part of *computational complexity theory* (CCT).

However, CCT is primarily concerned with the more general classification of problems and their inherent difficulty. According to CCT, tractable problems are those that may be solved by computer algorithms that run in polynomial time, i.e., the time (or number of steps) needed to find the solution is a polynomial function of the input. More explicitly, to say that a problem can be solved in polynomial time is to say that there exists an algorithm that, given an n-bit instance of the problem as input, can produce a solution in time $O(n^c)$, where c is a constant that depends on the problem. A problem is in the class P if it can be solved by an algorithm in polynomial time, and it is in the class NP (nondeterministic polynomial) if its solution can be guessed and verified

© Springer-Verlag GmbH Germany, part of Springer Nature 2018
R. Turner, *Computational Artifacts*, https://doi.org/10.1007/978-3-662-55565-1_23

in polynomial time. Intractable problems require times that are exponential functions of the input size.

If a problem is in NP, and all other NP problems are polynomial-time reducible to it, the problem is called NP-complete. Finding an efficient algorithm for any NP-complete problem implies that an efficient algorithm can be found for all such problems. Perhaps the most famous example of such a problem is the so-called *traveling salesman problem*. A traveler needs to visit all the cities in a list. The distances between all the cities are known. The salesman is only allowed to visit each city once, and must return to the starting city. The problem is to compute the shortest route. We can use a brute force approach to evaluate every possible tour, and then select the best one. If there are n vertices in a graph, there are $(n\text{-}1)!$ possibilities. Determining whether NP-complete problems are tractable or intractable is open, and one of the most important questions in theoretical computer science.

23.2 The Cobham-Edmonds Thesis

Turing's analysis of computation is often cited as a paradigm example of a case in which mathematical methods have been successfully employed to provide a precise analysis of an informal concept. Does the concept of a *feasible computation* admit of a similar analysis? According to the Cobham-Edmonds Thesis (named after Alan Cobham and Jack Edmonds) [40, 63],

> The complexity class P describes the class of *feasibly decidable problems.*.

In the paper "The intrinsic computational difficulty of functions" [40], Cobham suggests that P is a good way to describe the set of *feasibly computable problems*. Edmonds in his paper "Paths, trees, and flowers" [63] argues much the same.

Following the Church-Turing analysis of computation, one thing that needs to be demonstrated in support of the Cobham-Edmonds thesis is invariance. Different classes of machines need to yield the same classification of the inherent difficulty of problems. Here it is not enough to show that the formulations are extensionally equivalent, i.e., that problems solvable in polynomial time in one model are solvable in polynomial time in the others. One must also ensure that no residual complexity is hidden in the translations and simulations. It must be shown that the formulations can be translated without infeasible computational overheads. Fortunately, for a rather large class of models, such efficient translations exist. Van Emde Boas [64] formulated the invariance thesis thus:

> *Reasonable models of computation* can simulate each other within a polynomial bounded overhead in time and a constant-factor overhead in space.

However, from a practical perspective, there are some caveats. The Cobham-Edmonds thesis abstracts away from some crucial factors that affect the runtime. For example, it ignores the size of the exponent and the size of the input. There are problems in P requiring arbitrarily large exponents. A problem for which the best algorithm is $O(n^{100})$ is considered feasible by the Cobham-Edmonds thesis; on the other hand, a problem with an algorithm that takes 100^n is not. Actually, one could never solve an instance of size $n=2$ with the former algorithm, whereas an instance of the latter problem of size $n=2$ is solvable in real time.

Does the Cobham-Edmonds thesis challenge the Church-Turing thesis? It would now seem that formally computable functions include functions that are not informally computable – at least, not if informal computability means what a human computer can do with pencil and paper. We are much slower than computers, and they cannot feasibly compute more than the first two layers of the primitive recursive functions, i.e., the layers containing addition and multiplication. The *in principle* aspect of the characterization of computability is significant.

23.3 Quantum Computing and Feasibility

A final note on the long term prospects for the Cobham-Edmonds thesis: do quantum computers impinge upon this notion of feasible computation? There exists a quantum *solution* to an NP-complete problem, namely the satisfiability problem: given a proposition in the propositional calculus, the algorithm has to decide whether it has a satisfying truth assignment. Pitowsky [188] introduced a quantum algorithm that solves this problem by testing all possible assignments in a single step. Sounds promising. Unfortunately, any measurement performed on the output quantum state collapses it. Consequently, if there is one possible truth assignment that solves this decision problem, the probability of retrieving it is the same as in the case of a classical probabilistic Turing machine that guesses the solution and then checks it.

Pitowsky's conclusion is that in order to enhance computation with quantum mechanics we must construct superpositions that increase the probability of successfully retrieving the result: it has to be more successful than a pure guess. Shor's algorithm for factorizing large integers employs such a superposition. However, there appears to be some way to go before we need to worry about the security of our bank accounts – at least from this source.

Chapter 24
VARIETIES OF CORRECTNESS

When software does not work, does not conform to its specification, it is said to be *incorrect*; it is said to contain mistakes or bugs. Some instances are infamous. Mistakes in the software controlling the radiation therapy machine Therac-25 had fatal consequences, mistakes in the guidance system software of Ariane 5 caused it to crash, and a mistake in AT&T's software caused computer crashes. Although these are extreme cases, software and hardware errors are common. One way to quantify matters is in terms of the number of defects per lines of code [128]. In his book *Code Complete* [150], Steve McConnell states that the industry average is between 15 and 50 errors per 1000 lines of delivered code. But what is a *mistake*? When is a program taken to be incorrect?

There are two conceptually significant notions of correctness for computational artifacts. The first operates between the function and the structure: it is the correctness of the design. The structural description of the artifact must be in *accord* with the functional demands. This links two abstract structures. A second notion arises from the relationship between the structure and the artifact; it demands the correctness of the implementation. Here the artifact must be in accord with the structural demands, and it is here that the second notion, the one that links the abstract world with the physical, appears. On the face of it, the first notion is an abstract mathematical one, and the second an empirical one. It is with these two notions and their relationship that we shall be concerned, and here there is some controversy:

> The notion of program verification appears to trade upon an equivocation. Algorithms, as logical structures, are appropriate subjects for deductive verification. Programs, as causal models of those structures, are not. The success of program verification as a generally applicable and completely reliable method for guaranteeing program performance is not even a theoretical possibility [70].

Fetzer would have it that correctness proofs do not guarantee that the physical program, the one generated by the implementation, is correct. Even if we have correctness proofs for all the software involved in the implementation, we still have to deal with the abstract/physical interface. Programs running as physical devices are not mathematical things, and their correctness is not a

© Springer-Verlag GmbH Germany, part of Springer Nature 2018
R. Turner, *Computational Artifacts*, https://doi.org/10.1007/978-3-662-55565-1_24

mathematical affair. Correctness proofs link abstract objects, and do not guarantee that physical devices do not malfunction. However, some have been taken to disagree [114].

> When the correctness of a program, its compiler, and the hardware of the computer have all been established with mathematical certainty, it will be possible to place great reliance on the results of the program, and predict their properties with a confidence limited only by the reliability of the electronics.

But is there really any disagreement here? Is Hoare talking about the abstract program here, and Fetzer about the physical one? It would certainly seem so. In fact, Hoare himself alludes to this distinction in the very last part, where he talks about the reliability of the electronics.

Both seem to be agreed that at the level of the electronics, mathematical correctness is not even a theoretical possibility. If this is a correct analysis of their positions, it really is hard to see what all the fuss was about. All seem to be agreed that computational systems are at bottom physical things, and so there is always the possibility of unpredictable behavior. Both notions of correctness vary from artifact to artifact, and our task in this chapter is to catalog their natures by reference to our principal examples. This will prepare the ground for a more detailed analysis.

24.1 Digital Circuits

We begin with the simplest case, namely the case of digital circuits and logic machines.

Truth Table → Design → Digital Circuit → Implementation → Electronic Circuit

For simple Boolean circuits the correctness of the design, the digital circuit, amounts to the coincidence of its semantic definition with its functional specification. This is the correctness of the structural account relative to the functional description given by the truth table: the semantic truth table of the digital circuit must be in accord with the independently given truth table that lays out its functional specification. No physical devices are involved; this agreement is a formal or mathematical relationship, i.e., the two truth tables must be identical.

The correctness of the implementation is quite different. It will be correct if it generates a correct electronic circuit that agrees with the digital one. However, this is not a formal matter. While there may be a tight structural correspondence between the two, and this is formal in nature, this does not guarantee the correct physical behavior of the electronic circuit. This concerns the behavior of the physical device, not its form. The verification that the electronic circuit meets the demands of the digital one requires testing; it is an empirical matter. It is true that if the testing covers all cases we shall end up with a table that we require to be in extensional agreement with the table of the digital circuit, but getting to this point requires physical testing, not mathematical exploration.

This simple case clearly highlights the two different kinds of correctness that govern computational artifacts: the formal and the empirical.

24.2 Programs

High-level programs also have two forms of correctness that correspond to design correctness and implementation correctness. But here there are further complications. Consider the following two examples:

Z specification $\rightarrow$ *design* $\rightarrow$ Java programs $\rightarrow$ *implementation* $\rightarrow$ physical process
UML diagrams $\rightarrow$ *design* $\rightarrow$ Java programs $\rightarrow$ *implementation* $\rightarrow$ physical process

Again, the relations between function and structure are formal or mathematical in nature. However, in the first the relationship between Z and Java is more mathematically demanding. Here there must be agreement between both the type structure of the specification and the logical part contained in the predicate. In the case where the UML class structure acts as the functional vehicle, explicitly at least, there is only a demand for type or interface agreement. And generally, this is mechanically decidable. Of course, the UML specification can be beefed up to include logical demands added to those of the interface, but these do not form part of the standard UML class diagrams. So, although both cases are formal relationships, there is a difference in the implicit notion of correctness. In the UML case, unless it is supplemented with logical demands, we are only explicitly concerned with type or interface correctness.

Implementation correctness also generates greater complexity than that associated with digital circuits. It will still be the case that an individual physical device or program will be correct relative to a symbolic one if its physical behavior is in accord with the semantic interpretation of the symbolic program. In other words, its behavior on a physical machine must agree with its symbolic counterpart's behavior on the abstract machine. But the physical program is embedded in a complex weave of containing hardware and software, including layers of implementation and operating systems. The correctness of the physical program is established as a consequence of the correctness of the implementation of the whole language, and language implementations are complex artifacts. They involve compilers and interpreters, as well as the operating systems in which they are embedded. So, the correctness of the whole artifact is composed of several intermediate notions that involve its included compilers and the interpreters. The correctness of the former involves the correctness of the programs that are the compilers [144], and that of the latter ultimately involves the correctness of layers of devices through to electronic ones. Both mathematical and empirical notions are involved.

It is common practice to investigate and test a symbolic program via its implementation. For complexity reasons, this is often done in preference to its mathematical investigation. It is on the basis of the assumed correctness of the language implementations that we may employ the physical device to test the correctness of the symbolic one. Of course, in doing so we are in danger of building layers of incorrectness.

24.3 Software Systems

Finally, we consider software systems. Extracting the requirements in the first place is prone to errors. The model of requirements that is extracted might be wrong. Getting it right is an empirical enterprise that involves checking any proposed model against the demands of clients and stakeholders. The requirements specification might also be inconsistent in some formal sense, or the requirements may be physically impossible to satisfy. This resembles errors of theory or model building in science. Turing [228] introduces a similar distinction when he discusses errors of functioning and errors of conclusion. Errors of functioning would be mechanical or electrical faults that prevent a machine from doing what it is designed to do. In our terms, errors of conclusion involve mistakes in requirements elicitation.

However, once the engineer is satisfied that the model of the requirements is correct, it may be employed as a specification of the proposed system: the specification has governance over the structural design. We then have the possibility of structural-design incorrectness. The UML design then becomes a specification for the construction of actual programs, with all the above possible errors. And so on. As we move down through the levels of abstraction, more and more opportunities for disagreement between what is functioning as specification and what is functioning as structure, arise.

Fresco and Primiero [77] offer a classification of the type of errors that can be found in a software system. Their classification maps onto the distinctions drawn here. Material errors violate the correctness requirements of programs with respect to their specification, and performance errors arise when physical constraints are breached by some faulty implementing hardware. The design specification level concerns the correctness of the design. A design specification is, in its turn, instantiated in a high-level programming language – the algorithm design level. But these authors introduce more: the algorithm implementation level and the algorithm execution level. These correspond to the levels inside programs as artifacts. However, the difference between the formal and empirical notions of correctness will be our core focus.

Chapter 25
PROGRAM CORRECTNESS

The mathematical notion of correctness links the symbolic program with its specification, where correctness involves the construction of a proof that the program meets its specification. In this chapter, our aim is to assess the nature of such proofs. This raises several interrelated concerns about the kind of knowledge programmers have about their programs: is it traditional mathematical knowledge or some form of empirical knowledge? That this is not entirely clear emerges from several challenges to the simple, some would say simple-minded, mathematical picture:

- the mathematical challenge;
- the mechanical challenge;
- the pragmatic challenge;
- the scientific challenge.

These seek to undermine the basic claim that correctness proofs yield a form of mathematical knowledge. They raise epistemological concerns and, in turn, have methodological impact. They challenge the notion that correctness proofs are mathematical, and suggest that they are not even practically feasible, and consequently affect the very way that programs are taken to be linked with their specifications.

25.1 The Mathematical Challenge

One rather large fly in the mathematical ointment concerns the nature of the proofs involved in establishing properties such as correctness. In practice, in whatever system or method is employed, these proofs are long and tortuous to carry out. Consequently, some authors [52] claim that correctness proofs are very unlike proofs in mathematics. The latter are conceptually interesting and compelling, whereas in practice correctness proofs are long and shallow. In addition, they are normally carried out in formal systems. Typically, such proofs are dominated by numerous cases and

© Springer-Verlag GmbH Germany, part of Springer Nature 2018
R. Turner, *Computational Artifacts*, https://doi.org/10.1007/978-3-662-55565-1_25

shallow steps. In mathematical terms, they seldom contain much originality; they require patience rather than inspiration. Consider our proof of the correctness for the GCD. In many ways, this is untypical since the informal idea behind the proof is reasonably elegant. Indeed, it matches the elegance of the algorithm. However, even though it is for a very simple program, the actual formal proof is quite combinatorial. And matters get worse, far worse, as the programs approach industrial strength. This raises a question concerning the nature of such proofs: are they genuine mathematical proofs, and is the knowledge obtained from such proofs genuine mathematical knowledge?

To address these questions, we need to say a little about the nature of mathematical proofs. Unfortunately, here there is a fair amount of contemporary controversy. In the traditional picture of proofs in pure mathematics, notions such as *elegance* and *depth* are used to describe worthy ones. Although mathematical proofs may be very long and complicated, mathematicians construct proofs, definitions and abstract concepts in tandem. Mathematics is a creative process involving all these aspects. In particular, the process of proving complex theorems may involve the invention of new mathematical concepts and notation. They may even await the invention of new areas of mathematics.

One characteristic of mathematical proofs, which is often taken as part of their essential nature, concerns their ability to be grasped as a whole. While they may require study and reflection, genuine mathematical proofs must be such that they may be *taken in all at once*. Such proofs are referred to by Hacking [100] as *Cartesian proofs*. For Descartes, reasoning is self-authenticating only if you can grasp an entire proof; he speaks of getting an entire proof in the mind all at once. A related aspect has it that a proof must not merely convince us that a result is true, but must explain to us why the result is true. Mathematical proofs involve abstraction and generalization. They should be eventually clear and obvious, at least to the appropriate mathematical community, and, by logical standards, they may be quite informal.

Wittgenstein also appears to favor such proofs when he asserts that proofs must be surveyable [256]. For him, *graspable*, when applied to an axiom or whole proof allows it to be used as a rule. Proofs have a normative function, and the ability to grasp their content is a prerequisite for using them as rules.

There is a great deal more to be said about these notions, but what is clear is that, in some form or another, many leading mathematicians see such proofs as the central ones of mathematics. In particular, Grothendieck [96] puts things in the following way:

> A vision that decants little by little over months and years, bringing to light the "obvious" thing that no one had seen, taking form in an "obvious" assertion of which no one had dreamed ... and that the first one to come along can then prove, in five minutes, using techniques ready to hand.

Genuine proofs should make everything explicit and obvious. Presumably, this makes them graspable. Mathematical creation is a vision that, little by little, brings to light the obvious. The important notion is not an individual proof but a proof idea; such proofs involve techniques of proof that rapidly generalize. Mathematical creation involves a vision, a new idea, etc.

Whatever the merits and clarity of this picture, correctness proofs seem not to fit it. Formal correctness proofs do not seem to involve the creation of new concepts, let alone new areas of mathematics. They have little conceptual significance and do not bring about conceptual change and display connections between existing concepts. This is the core of the mathematical challenge: large, conceptually shallow proofs are not genuine mathematical proofs. Consequently, they do not deliver genuine mathematical knowledge.

However, there is a different notion of proof operating in contemporary mathematics [100]. In contrast to Descartes, Leibniz discusses a notion of proof in which every step is meticulously laid out, and can be checked, line by line, in a mechanical way. Such proofs are close to the kind of formal proofs found in formal logic. They are arguably safer in terms of their veracity since each step is minimal and easy to check. They may even be mechanically checked by the appropriate software. Moreover, some claim that mathematics is becoming more like this. A prominent mathematician from this camp, Voevodsky [242] believed that the most interesting and important proofs in current mathematics are like the Leibnizian ideal. He further claimed that we are entering a new era that will be characterized by the widespread use of automated tools for proof construction and verification.

Mathematics at the computer: mathematics carried out with the aid of computer proof assistants. Soon, mathematicians will not consider a theorem proven until a computer has verified it.

According to this perspective, correctness proofs not only qualify as mathematical proofs, but they are part of a new era of mathematics. Correctness proofs seem to have been reinstated as deliverers of mathematical knowledge.

Everything depends upon which of these perspectives one finds the more attractive or persuasive. However, while I suspect that most pure mathematicians still favor the Cartesian perspective [106], the alternative is gaining ground because of the complexity of modern mathematics. Unfortunately, even if correctness proofs are seen through Leibnizian spectacles, there is yet another challenge to their epistemological status.

25.2 The Mechanical Challenge

Because of their size, involving the checking of untold cases, these proofs are rarely carried out by hand. In practice, they are constructed with the aid of theorem-provers. But, in the wake of this, come some concerns.

When theorem-provers are employed, correctness proofs are derived by representing the programs as axiomatic theories, and their specifications deduced as consequences of those theories. More sophisticated systems employ model checking [15], where a proof of correctness is obtained by an algorithm that checks whether the program is a model of the specification. While this may reduce the correctness problem to that of a single program, it still means that we are left with the correctness problem for a program. So, we have replaced the correctness problem for one program by another, and we have the beginning of an infinite regress.

The use of theorem-proving techniques to construct proofs brings a physical device into the arena of mathematical proofs. And this raises the following question: are they mathematical proofs when they rely on a physical device? Computational systems are at bottom physical systems. theorem-provers are implemented programs, and their results depend upon a physical computation. When theorem-provers are used, the results only yield knowledge on the assumption that the underlying physical devices are correct. So, in showing that a mathematical proposition is true, at some level, we need to show that some physical machine operations meet their specification. And this involves empirical verification. In the limit, theorem-proving systems run on physical machines, and so the software is only correct if these mechanical systems are. Ultimately, such correctness amounts to empirical verification. If this argument has substance, in practice, the correctness of software is an empirical enterprise, not a mathematical one.

Of course, one might claim that even if we use pencil and paper to do proofs by hand, we are using physical devices in the construction of mathematical proofs. But there is a difference. In the case of computers, physical devices are carrying out the proofs; in the case of pencils we are. Pencils are an aid in our construction of proofs. In contrast, as Annie Lennox might put it: "computers are doin' it for themselves" .

In this regard, one might demand that correctness proofs be constructed by hand, but be checked by a computer. Indeed, as we have noted, Voevodsky saw this as a necessity for contemporary proofs. Of course, the proof checker is itself in need of checking. Arkoudas and Bringsjord [12] argue that since there is only one correctness proof that needs to be checked, namely that of the proof checker itself, then the possibility of mistakes is significantly reduced. But this is beside the point from the perspective of the mechanical challenge: if the proof has been constructed not by a computer but by a human, then there is no objection to this step – since it is in addition to the traditional construction of the proof.

Mathematical knowledge is generally taken to be a priori knowledge. Burge [31] argues that knowledge of such computer proofs can be taken as a priori knowledge. According to him, *a priori* knowledge does not depend for its justification on any sensory experience. However, he allows that such knowledge may depend for its possibility on sensory experience; for example, knowledge that red is a color may be a priori even though having this knowledge requires having sensory experience of red in order to have the concepts required to even formulate the idea. If correct, this closes the gap between a priori and a posteriori claims about computer-assisted correctness proofs. This is now an issue at the heart of the contemporary philosophy of mathematics, where the use of computers seeks to change our very notion of mathematical knowledge.

25.3 The Pragmatic Challenge

We have already alluded to this in an earlier chapter, but here we consider the epistemological implications. Within the appropriate mathematical framework, proving the correctness of any linguistic program, relative to its specification, is theoretically possible. However, real software is big and complex. In such cases proving correctness is generally taken to be practically infeasible. In practice, these proofs are rarely carried out, with or without the aid of theorem-provers. In practice, proofs are replaced with testing and verification. From a methodological perspective, the relationships between program and specification is taken to be an empirical one to be uncovered by testing and verification rather than by any form of proof. This is the *pragmatic* challenge.

The complexity of modern software has led many practitioners to be skeptical about the mathematical approach to software development. Programs often contain tens of thousands of lines of code. Even Hoare acknowledges that in practice the construction of correctness proofs is rare, but he blames this on the education of programmers and language designers [115].

Consequently, instead of proofs, programs are run with known results to check that they give the right answers, agree with the specification. Instead of detailed formal proofs, judiciously designed tests are employed to examine whether the program meets its specification. Of course, this assumes that we know some of the correct answers, i.e., we know from the specification what the results should be. A physical system is being used to test the *correctness* of a mathematical object. Correct results for known instances inform us that the physical program is working according to the specification. What we conclude is an empirical observation about the physical program. On the face of it, the testing relates the physical program and the specification and bypasses the symbolic one. However, on the assumption that the implementation is correct, correct results for the physical program tell us that the symbolic program meets its specification – at least on the test samples. The physical program is seen as a window on the symbolic program which is being indirectly tested via the physical system. This provides a means of sidestepping or avoiding the construction of a correctness proof. If the implementation is correct, then the physical program and the symbolic one will be in harmony. So, we may indirectly test the correctness of the symbolic program via its physical counterpart. In the process, there is no direct appeal to the semantics of the program. This only gets traction via the correctness of the implementation. But now we are faced with the correctness of the implementation. In practice, this too is subject to testing and verification rather than any notion of mathematical correctness. Indeed, at bottom level, the implementation is a physical device that is not amenable to mathematical proof.

What kind of knowledge does this return? And what is it knowledge of? One thing is clear: it does not provide any guarantee of mathematical correctness for the symbolic program:

Program testing can be used to show the presence of bugs, but never to show their absence! [61]

Testing and verification seems only to yield empirical evidence. Prima facie, it is direct evidence that the physical program meets the specification; indirectly, it is taken as indirect evidence that the symbolic program does. Once more, programs are not treated as mathematical entities, and our knowledge of them is not mathematical knowledge.

25.4 The Scientific Challenge

Given this approach to correctness, it would appear that software development is a scientific activity that involves simulation and experimentation [78]. Consequently, our knowledge of our programs is scientific knowledge. Indeed, there is a hint of such a perspective in many of the Turing Award lectures [126, 230], where one can find many variations on the more general claim that computer science is a scientific discipline. But matters are somewhat more involved. While these methods may deliver some form of empirical knowledge, it is unclear that it is of the same kind as that involved in the empirical testing of scientific theories.

Many of the experiments performed by software engineers are aimed at exploring the actual functioning of an artifact [199]. Experiments are performed to evaluate the behavior of a system, and to examine its capabilities and its comparative properties. These experiments do not directly link a program with its specification. They are exploratory experiments that appear to introduce new styles of experiment. Tedre [222] provides an excellent study of the different notions of experiment to be found in computer science.

However, even when the main aim of such experiments is to establish the correctness of a program or system relative to its specification, we are prompted to ask in what sense this testing is aimed at the verification of a model or theory, as it would be in mainstream science. When a program is tested, it is tested against a specification. Specifications are not models of anything. Models are descriptive, and specifications normative. Consider the design of a teapot. The abstract notion of a teapot is not a model of what an artifact, the teapot, does. If it were so intended, and the teapot did not brew and pour properly, then we would change the model to meet the properties of the teapot. But this is precisely the opposite of practice. If the teapot does not brew and pour properly, we change the teapot, not the model. In science, it is the physical phenomena that provide the evidence for the testing of a model. If the physical thing does not conform, we change the model. While program verification employs testing and verification, its purpose is not the verification of a model or theory in the sense of empirical science. When a program is tested, it is tested against a specification. If it does not meet its specification, we don't change the specification, we change the device.

In summary, there is a fundamental difference in the underlying logic of specifications versus models/theories. There is a methodological distinction: a falsifying test leads to the revision of the artifact, not of the hypotheses as in the case of a scientific hypotheses. This is due to the difference in the intentional stance of specifications and empirical hypotheses in science [235].

This is not to say that software testing and verification do not employ *model-based* techniques. Often, the software system is modeled because of the complexity of the real thing. Axiomatic systems and state transition systems are used to evaluate whether the computational artifacts conform or do not conform to their specifications. These can be understood as theories of the represented systems in that they are used to predict and explain the future behavior of those systems. For example, in [10] Angius and Tamburrini convincingly argue that the mathematical structures used in model checking are in compliance with Suppe's definition of scientific models. This stage of correctness involves the construction of an empirical model that takes us even further away from the pure mathematical picture. Here there is a direct analogy with scientific models. If the model gets things wrong i.e., does not capture the system, then it will be modified or replaced. This process resembles standard model construction in science. This is yet another shift in perspective that seemingly takes us closer to the notion that establishing correctness is a scientific affair.

But still, ultimately what is being tested is the artifact, not any notion of a model. The methodology may be similar, but we are dealing with a designed artifact. The function of any model is to stand proxy for the system in the process of verification. We are not trying to model the system as a natural thing, but as a technical artifact. The overall use of the model is to demonstrate the correctness of the system, it is not to model it as a natural device. The normative role of the specification is still in force, except that now the specification has normative force over the system indirectly via the model. On the face of it, because computer scientists build mathematical models of complex computational systems, it does not mean that they are engaged in a scientific activity in the same way that physicists are. Ultimately, they are still involved in creating technical artifacts not constructing models or theories about natural things.

25.5 Proof and Principles of Program Design

The principles of modularity and abstraction aim at reducing complexity in program and software development. These are principles of simplicity, and are closely allied with the themes that underlie Cartesian proofs. Both involve comprehension and seek to avoid complexity. These programming principles aim to generate simple programs, and presumably simple programs lead to simpler proofs of correctness. Indeed, the rigorous application of these principles should lead to programs whose correctness is almost immediate. But there is little evidence that this is what happens in practice. Single programs often still run into hundreds of lines of code. Perhaps the programmers are not employing these design principles with sufficient rigor and thoroughness. But whatever the truth of the matter is, what seems clear is that in practice programmers do not have mathematical knowledge of their creations.

Chapter 26
TYPES AND CORRECTNESS

While in practice full correctness proofs are rarely forthcoming, computer scientists still employ tools that guarantee some form of correctness. One of these tools is a type checker for the language. Types fine-tune the grammatical structure of the language, and bring semantic intuitions to the grammatical party. The following is a quote from Pierce's text on types in programming languages. It makes it clear that the central role of types is correctness, and that type checking is a step towards correctness:

> Despite decades of concern in both industry and academia, expensive failures of large software projects are common. Proposed approaches to improving software quality include – among other ideas – a broad spectrum of techniques for helping ensure that a software system behaves correctly with respect to some specification, implicit or explicit, of its desired behavior. On one end of this spectrum are powerful frameworks such as algebraic specification languages, modal logics, and denotational semantics; these can be used to express very general correctness properties but are cumbersome to use and demand significant involvement by the programmer not only in the application domain but also in the formal subtleties of the framework itself. At the other end are techniques of much more limited power – so limited that they can be built into compilers or linkers and thus "applied" even by programmers unfamiliar with the underlying theories. Such methods often take the form of type systems.

At one end of the spectrum, types are analogous to dimensional analysis in physics, and here type checking is decidable; at the other end there are systems of types such as those of constructive type theory [156] where types serve as specifications, and proofs that a type is inhabited generate correct programs for that type. However, in these latter systems the construction of the proof is a creative enterprise parallel to proving correctness after the program has been constructed. Consequently, designing a type system is an optimization problem: one must tread a path between total correctness and decidability. Any type system that delivers the total correctness of programs will not be decidable, since program termination is not decidable. On the other hand, decidability of the type system allows the process of type checking to be built into the implementation of the

© Springer-Verlag GmbH Germany, part of Springer Nature 2018
R. Turner, *Computational Artifacts*, https://doi.org/10.1007/978-3-662-55565-1_26

language. And there are variations on how this is done. Our main objective is to clarify these various options, and to evaluate the contribution of type checking to the establishment of correct programs.

26.1 Type Inference

There is a distinction between the type checking program that forms part of any implementation, and the actual system of type inference. The latter provides the functional specification of the former. To illustrate matters, we employ a simple language of expressions with two basic types: numbers and Booleans:

$E ::= x \mid 0 \mid 1 \mid E{+}E \mid E\mathbf{x}E \mid \text{true} \mid \text{false} \mid E{<}E \mid \neg B \mid B{\wedge}B \mid$

$T ::= N \mid \text{Bool}$

Independent of the mechanism of type checking are the logical rules that constitute the type inference system itself. This determines what it is to be well-typed. In this system, variables obtain their types from a type context. Such contexts (d) are either individual type assignments $(x : T)$ or sequences of such $(x_1 : T_1, x_2 : T_2, \ldots, x_n : T_n)$. Relative to such type assignment contexts, expressions receive their types via *rules of type inference* such as the following:

$$\frac{d \vdash E_1 : N \quad d \vdash E_2 : N}{d \vdash E_1 + E_2 : N} \qquad \frac{d \vdash E_1 : N \quad d \vdash E_2 : N}{d \vdash E_1 < E_2 : Bool}.$$

The first rule insists that addition can only be applied to those expressions that type check as numbers. This is a grammatical rule that encodes the semantic information that numbers may be added together. The second insists that numbers can be compared relative to the less than relation, and the result is a Boolean. There are similar rules for the other arithmetic and Boolean connectives, and some simple housekeeping rules for the contexts.

Now consider the following. Given the context $x : N; y : N$,

$$x + y$$

is *well-typed* since, according to the first rule, we may add two numbers. However, we would not normally add Boolean values. So, given the context $x : Bool, y : Bool$, the following is not well-typed.

$$x + y$$

This is not sanctioned by the rules. Rules such as these form the basis of type checking algorithms that live somewhere in the implementation process. There are all sorts of caveats here, but the general picture is clear enough: when we perform operations on data items, the type discipline needs to be respected. The type rules impose a little semantic structure on the world of data. This is for the user's benefit: types force a conceptual framework on the programmer, and the type checker helps her to obey the principles imposed by it.

26.2 Semantics

Type systems are motivated by semantic considerations: with our normal understanding of numbers we can add two numbers, but with our normal understanding of Booleans we cannot. How are the semantics and the type system related? In what follows we shall assume that the state and the type context are consistent, i.e., the value the state s assigns to each variable x must have the same type as that assigned to x by the type context d. We may then state the *soundness* and *completeness* properties of the type system relative to the semantics:

- The type system is *sound* relative to the semantics if, for every expression E, if $d \vdash E : T$, then $< E, s > \Downarrow v$, where $d \vdash v : T$.
- The type system is *complete* relative to the semantics if, for every expression E, if $< E, s > \Downarrow v$ where $d \vdash v : T$, then $d \vdash E : T$.

Soundness ensures that the type system agrees with the semantics in that, if the type system assigns a type to an expression, then it is assigned a corresponding value under the rules of evaluation, i.e., by the operational semantics. Completeness is the converse: if an expression has a semantic value of a certain type, then it will be assigned that type by the type system. Soundness ensures that the type system is sanctioned by the semantics, and completeness guarantees that the type system does everything that is semantically warranted.

How does incompleteness come about? The semantics might go beyond the type system and assign values to cases that do not type check. For instance, it might evaluate addition by coercion:

$$\frac{< E, s > \Downarrow v \qquad < E', s > \Downarrow v'}{< E + E', s > \Downarrow \mathrm{coerce}(v) + \mathrm{coerce}(v')}$$

where $\mathrm{coerce}(a)$ returns a if a is a number, 0 if it is the Boolean value false, and 1 if it is the Boolean value true. The type system with coercion is sound with respect to this semantics, but it is not complete. The expression *true+false* will get a numerical value 1 in the semantics, and hence will be of type N, but will not be assigned a type of any description by the type system.

However, we may complete matters. We add a new expression to the grammar *coerce(E)* and new rules to the type system:

$$\frac{d \vdash E : Bool}{d \vdash \text{coerce}(E) : N} \qquad \frac{d \vdash E : N}{d \vdash \text{coerce}(E) : N}.$$

If we now semantically identify the semantic coercion as the meaning of the syntactic operation, then the type system is again complete. A minimal requirement on a type system is that it be sound. Complete type systems are often called *strong*, and incomplete ones *weak*. Intuitively, Java is a strongly typed language, whereas Perl is not. So, the notions of strong and weak refer to the logical strength of the type system relative to its semantics.

Strong typing is closely associated with correctness and security, since simple semantic errors concerned with type mismatch will be picked up. There are no surprises: what you see is what you get. The programmer has complete information, and nothing is hidden in the implementation. Unfortunately, there are many examples of languages that allow hidden coercions in the implementation. If these are hidden from the user, correctness is challenged.

26.3 Implementation

Although coercion can happen within compiled languages, it is more likely to occur in interpreted languages or in dynamically typed languages i.e., languages where type checking is carried out during the execution of the program. Indeed, many discussions of strong and weak typing are cast in terms of the implementation, i.e., within the type checking system itself. But where in the implementation is the type checking done? There are two basic options:

- compile time;
- run time.

Which is best? Language designers are split on the issue. This is clear from a discussion [22] between Bill Venners and Josh Bloch defending their approaches to typing.:

BV: In your book you say, it is always beneficial to detect programming errors as quickly as possible. I've met people who don't feel that way: people from the Smalltalk community, people who like Python, and so on. These people feel that all those compile time errors get in the way of their productivity. They feel more productive in a weakly typed environment, where more problems must be discovered at runtime. These people feel that their weakly-typed language of choice gives them as much robustness, but more quickly, than strongly-typed languages such as Java.

JB: I quibble with the fact that they are getting as much robustness. I suppose the extreme example of that is shell scripts, which are interpreted. There is no compile time. You can code anything you want. And I think anyone who has used shell scripts has seen them blow up in the field. In fact, people don't expect them to run on all inputs. If you take a shell script, try to do something fancy with it, and it doesn't work, you say Oh well, I guess it doesn't handle that. And you play around with the inputs and try to find something it does handle. There's no doubt that you can prototype more quickly in an environment that lets you get away with murder at compile time, but I do think the resulting programs are less robust. I think that to get the most robust programs, you want to do as much static type checking as possible. I do understand that people coming from these environments really do find static type checking constraining. It's no fun to deal with compile-time errors, but there are real benefits in terms of robustness. So, you pay your money and you take your choice.

Seemingly, there is a conflict between correctness and efficiency: compile-time type checking is claimed to be more secure, but less efficient in the development of code. This is a dispute over the appropriate methods of programming, and how they deliver correctness and robustness. Compile-time type checking is taken to be more secure since errors are picked up early. Once again, we have a clash between correctness and efficiency.

Notice that BV is using the phrase *weakly typed environment* to refer to run-time checking, and JB responds with the phrase *static type checking*. There is considerable ambiguity in the terms used in the literature. But here the distinction refers to the implementation; it is a distinction between compile-time checking and run-time checking. It is not the above distinction of (semantic) weak and strong typing.

However, one can address the same issue in terms of the semantics. If we view the abstract semantics as an abstract interpreter, then we have two options: we could type check before the semantic function is applied, and only assign a meaning to well-typed terms, or we could apply the semantics freely but only accept outputs that are well typed. This is the semantic version of compile-time versus run-time checking. So, we have a parallel distinction in the semantics and implementation.

26.4 Correctness

Even with a complete compile-time type system, we have no guarantee that programs will meet their specifications. To illustrate the difference, consider addition and multiplication. They both have the same type: they input two numbers and output one. But their specifications are different. Showing that a program is of the given type does not show that it satisfies a specification of addition or multiplication. For a second example, consider the implementation of sets as lists. The type of union on sets is the same as that of intersection. From a type-theoretical perspective, they could both be implemented as list concatenation. The implementer is free to implement the operations and objects with any objects of the appropriate representational type. For example, she could, without breaking the rules of the type systems, implement set-theoretic union and intersection the same way. In general, type declarations do not constitute complete functional specifications. The matching of types is clearly not enough for correctness.

So, how do matters proceed in practice? If we are told we are dealing with sets and their operations, then any implementer who knows or can uncover their import already has more information than is present in the type declarations. This information is implicitly taken to be part of the functional demands, and used as part of the program design process. The full functional specification is implicit in the background knowledge of the implementer.

What is less clear is what happens when the system under construction is unfamiliar to the programmer. Imagine the module is a system where the signature includes operations named *throw* and *catch*. Names may well serve as metaphors. For instance, the English words *throw* and *catch* conjure up images that are associated with games with balls. As such, they may guide the program designer to more precise requirements. But, equally, such guidance may well be too vague and ambiguous for use in practice. That is the nature of metaphor and analogy. In such a case, this information must be supplemented. This may also include advice from the client. Whichever route is taken, and whatever method is employed, such maneuvers must somehow result in an actual functional specification. However the program designer proceeds, the fact remains that, by itself, a decidable type system cannot provide a functional specification. At some point in the design process some more demanding functional specification must be produced. Even though they have the same type definition, set intersection is not the same as set union.

However, this is a very precise example and might be taken to be untypical. In our discussion of precision and information, we suggested that something could be precise relative to a given level of abstraction. If we apply this to our library system, then matters are murkier. Suppose the operation of returning a book to the library is given the type

$$Patron \otimes Book \Rightarrow (Library \Rightarrow Library).$$

So, given a reader and a book, it changes the state of the library. Is there anything else we need to specify about this operation aside from its type? Yes, we need to know what it is supposed to do. How does this differ from the operation of borrowing a book, which would seem to have the same type? But can this information be packed into the type? With a richer system of types we could express the difference between these two operations. For example, suppose that we allow sub-types in the type system:

$$\{x : T | \phi\}$$

Here ϕ is a well-formed formula of some logical system such as predicate logic. This would enable the *Library* type to contain more information about which books are on loan to which readers. However, as we said at the outset, there is a trade-off: enrich the type system too much and decidability is sacrificed. This imposes proof obligations on the programmer.

Chapter 27
THE SIMPLE MAPPING ACCOUNT

We now move on to the second notion of correctness, that which governs physical devices. On the face of it, this raises a very different kind of conceptual problem from the mathematical one. Electronic devices fail for a variety of reasons: mechanical impact, excessive temperature, or very high voltage may all cause failure. How do we test such circuits for correctness? We have already indicated that some notion of verification is involved, but what form does it take? Here we are not referring to the use of formal techniques for establishing the correctness of the abstract digital circuit against its truth table specification [32]. Nor are we referring to the use of formal methods to ensure the correctness of hardware designs, where specification frameworks involving temporal logics and algebraic methods are employed. We are referring to the verification of the electronic device by empirical testing. While this may employ many different means and methods, such as model checking, algebraic automata and theorem proving, in the end it must involve empirical work. But what is the goal of this work; what does correctness amount to?

27.1 The Correctness of Electronic Devices

To make matters more concrete, consider an electronic circuit and a truth table specification. What constitutes agreement between the two, or, rather, when does the physical device behave in accord with its functional specification? The standard answer involves the input/output behavior of the physical device: agreement demands that the input/output behavior of the electronic device, its physical behavior, is, via the voltage interpretation, identical to that demanded by the truth table.

In general, the core of verification is functional validation and involves the production of large suites of standalone tests. Somehow, the functional model of a design is simulated with typical input, and the output is then checked for the expected behavior. Unfortunately, with current methods [7], more than 70% of design time and engineering resources are spent on verification. So, the correctness

© Springer-Verlag GmbH Germany, part of Springer Nature 2018

R. Turner, *Computational Artifacts*, https://doi.org/10.1007/978-3-662-55565-1_27

of electronic circuits is a bottleneck for program correctness. As Hoare puts it, our confidence in any software is *limited only by the reliability of the electronics*. It appears that this confidence should not be too high. But this is not the focus of this chapter, which is the notion of correctness itself: the notion of agreement between the functional specification of the circuit and the electronic device.

27.2 Simple Mapping Account

The simple mapping account of physical computation is due to Hilary Putnam [190]. Roughly, anything that may be described by a physical computation is determined by an abstract one. But what is it to fix a physical system via an abstract one? For the sake of concreteness, suppose that the abstract machine is a Turing machine where a Turing machine table consists of one column for each of the (finitely many) internal states of the machine, and one row for each of the machine's symbol types. The machines table specifies what the machine does for each symbol and internal state. A physical computation system is one in which the *corresponding* physical states and operations mirror the corresponding abstract ones. This would lead to a characterization of a physical computation as being an image of a Turing one.

However, this approach to physical computation does something more than define the notion of physical computation. Implicitly, it employs an account of *physical implementation*. It tells us what the latter is taken to be: a physical *implementation* of an abstract system of states and operations is a physical one that mirrors the abstract one. In setting out the idea of a correspondence we are actually defining the notion of *implementation*. It is this application of Putnam's idea of computation that we shall be concerned with here.

A more generic approach that abandons Turing machines is what Godfrey-Smith [89] refers to as the *simple mapping account* of computation. According to this account, a physical system P performs a computation A just if and only if the following holds:

1. There is a mapping from the states of the physical system P to the states of an abstract system A, such that the following is true.
2. The state transitions between the physical states mirror the state transitions between the computational states: for any abstract state transition $s_1 \Rightarrow s_2$, if the system is in the physical state that maps onto s_1, it then goes into the physical state that maps onto s_2.

What happens if we apply this to our digital circuit? Under the simple mapping account, a physical implementation of our circuit is any physical system for which there is the appropriate mapping between its states and the states of the truth table. Notice that the following conditional is taken to be the material conditional of standard propositional logic:

If the system is in the physical state that maps onto s_1, it then goes into the physical state that maps onto s_2.

It requires only that the consequent be true when the antecedent is. Consequently, we only seem to require that the artifact relation be one of extensional agreement, i.e., be in agreement with the abstract system. This is precisely what we have already demanded for the correctness of our electronic circuit. In other words, the correctness of our digital circuit is an instance of the simple mapping account. But put in this way, this opens the floodgates to other possible implementations. In particular, the following physical setup, where the states of the machine are represented as tokens painted white or black and arranged as follows, would be sufficient:

Input	Input	Sum	Carry
white	white	white	white
white	black	white	black
black	white	white	black
black	black	white	black

The physical state transitions are extensionally fixed by the table, and these are all that is required by the simple mapping account. Consequently, a physical copy of the original functional specification will, under the simple mapping account, serve as an implementation. Moreover, there is nothing to prevent the replacing of this table by another that has the same external structure. It appears that such mappings are relatively easy to come by. It is this notion of implementation that leads some authors to conclude that almost every physical system (as long as it has enough parts) implements almost every abstract one. It is this puzzle that we shall be concerned with in this chapter.

Notice that to set up the correspondence in the first place, we must compute the whole state table for our abstract machine, which means that the human computer is doing all the computation. Of course, in general this will be practically impossible. The relationship is created post hoc. This point is made by Copeland [45], who observes that the mappings of the simple mapping account are illegitimate because they are constructed after the computation is already given. One might claim that, in itself, the physical table provides no method of computation. Unfortunately, it does. It is just a very bad one that does not explain how the table was constructed.

On the face of it, the simple mapping account is only concerned with the formal relationship between the extensional structures of the two devices. Of course, in general, not only can we not compute the whole extensional table, but doing so also defeats the whole point of building a physical computer. Be that as it may, this move does not address the fundamental concern raised by the simple mapping account which seems not to sufficiently constrain the notion of implementation.

Moreover, the simple mapping account does not just infect the physical case of implementation; we shall now argue that, extensional agreement or regularity also applies to the mathematical notion of correctness.

27.3 The Mathematical Case

Consider again the following specification of the square root for natural numbers:

$$SQRT(x : Real, y : Real)$$

$$Pre : x \geq 0$$

$$Post : y * y = x \wedge y \geq 0$$

SQRT defines a relation between natural numbers and real numbers where *Num* and *Real* are the data types for the specification language. In actual practice, we will always be dealing with finite sets and types and finite operations on them, although this is not essential to the underlying argument in relation to the mathematical case.

Now consider an abstract table that associates the square root with the numbers $1, 2, 3, 4, 5$ which, for the sake of illustration, we take to be all there are in the data type *Num* of the specification language.

Input	1	2	3	4	5
Output	1	1.41421356237	1.73205080757	2	2.2360679775

For this table, we then have

$$\forall x : Num.\forall y : Real \cdot x \geq 0 \rightarrow (Abstracttable(x, y) \rightarrow SQRT(x, y))$$

It would seem that, under extensional agreement, the abstract table satisfies the specification. For the more general case, suppose that the type of reals of the programming language is different. Suppose it takes the following form: $Real = \{a, b, c, d, e\}$. This may seem a little odd, but it is perfectly plausible that a programming-language representation of common mathematical domains may be nonstandard. To link the two, we must relate the classes of reals. We interpret the programming class in the specification language as

$$I(a) = 1, \quad I(b) = 1.41421356237,$$

$$I(c) = 1.73205080757,$$

$$I(d) = 2, \quad I(e) = 2.2360679775.$$

This now applies to all computations over the reals; it is the general way that the programming language is interpreted in the specification language. For our square root program, the statement of correctness now takes the following more general form:

$$\forall x : Num.\forall y : Real \cdot x \geqq 0 \rightarrow (Abstracttable(x, y) \rightarrow SQRT(I(x), I(y)))$$

where the abstract table that is our program, is now given as follows.

Input	1	2	3	4	5
Output	a	b	c	d	e

Under the given interpretation, the above table implements the square root. In other words, we can always find an interpretation such that any table of the right shape and size is a correct implementation of any specification.

So, we have a similar problem at the mathematical level; apparently, even abstract mathematical correctness amounts to little more than extensional agreement. At every level of abstraction, correctness reduces to extensional agreement, and, ontologically, extensional agreement is too easy to come by.

27.4 Good and Bad Programs

One thing that the physical and mathematical cases have in common is badness. In so far as they satisfy their functional specifications, in both cases the extensional solutions are solutions. However, they are bad solutions. This is particularly clear in the mathematical case. Recall the solution based upon Newton's method to find the square root of a positive number n,

$$x_{k+1} = \frac{1}{2} \left(x_k + \frac{n}{x_k} \right)$$

At each stage the new value is computed from the old one, and the process continues until the old and new values (x_k and x_{k+1}) converge. We do not need to look at any code to see why this is a better solution than just listing the table of values. We are not appealing to any notion of computation here: we are not suggesting that Newton's method is an algorithm and the extensional solution is not. In fact, they are both algorithms. But we are suggesting that one algorithm (and any corresponding program) is better than the other. Why? Any answer to this must refer back to the discussion of good design and good programs. For example, the program that implements the extensional solution consists entirely of case statements:

```
Case  x=1  then  y=1
Case  x=2  then  y=1.41421356237
Case  x=3  then  y=1.73205080757
Case  x=4  then  y=2
Case  x=5  then  y=2.2360679775
```

Any student asked to write a program for the square root would get very few marks for this solution. It is not uniform: the solution is just a collection of case statements. It is not generic: it is restricted to a given set of values. It is not transparent, in that it is not clearly a solution. Indeed, a proof that it is a solution would involve carrying out the computations, and checking the table entry by entry. One reason that the Newtonian solution is a better one is that it is uniform over a whole range of values. Another is that it is explanatory: the algorithm returns the square root of n since, when the two values of x converge we obtain a quadratic equation for x whose solution is the square root of n. In general, insisting on good and explanatory solutions cuts down the options. Likewise, in the physical case, good solutions should involve simple, elegant, and explanatory digital designs that, for instance, do not just enumerate all the possible outcomes in an electronic table.

In particular, both solutions (the physical and mathematical cases) appeal to some notion of explanation, even if rather different ones.

Chapter 28
COMPUTATIONAL EXPLANATION

A design also contains (at least implicitly) an explanation of how the proposed physical system will be able to perform the required function. In other words, a design also consists of a technological explanation, i.e., an explanation of the function of a technological object in terms of the physical structure of that object. A technological explanation is an integral part of a design and plays a crucial role in justifying a design: it shows that on the basis of it's physical structure an object will perform a certain function [139].

A design contains an explanation of how an artifact realizes its function. But, how exactly? In particular, how are programmers and software engineers able to predict and explain the outcomes of their software designs before they have been implemented?

In this chapter, we provide an overview of computational explanation. We are not asking how computational mechanisms explain all physical phenomena. We are not concerned with the general use of computational models in science. Our interests are much more limited: we are only concerned with how computer scientists are able to explain why their designs satisfy their specifications. Once more, we shall address this by reference to our three familiar examples of computational artifacts, machines, programs, and software systems, where, on the face of it, three rather different notions of explanation seem to be in play.

28.1 Technological Explanation

The simple mapping account has driven some authors to attempt to provide an account of implementation that somehow restricts the class of possible extensional interpretations. One proposed solution demands a causal connection between the components of the physical system [35, 45]. Consider again the statement of physical correctness or agreement. Here, the use of the material conditional is said to be the hub of the problem. Consequently, the way forward is to interpret the conditional counterfactually rather than materially, i.e., we are to interpret the following counterfactually:

© Springer-Verlag GmbH Germany, part of Springer Nature 2018
R. Turner, *Computational Artifacts*, https://doi.org/10.1007/978-3-662-55565-1_28

If the system is in the physical state that maps onto s_1, it then goes into the physical state that maps onto s_2.

Apparently, it is the lack of causal connection between the states of the physical device that needs to be addressed. The counterfactual interpretation of the conditional ensures that the mechanism causally mirrors the abstract one. While there are some standard objections [210] there is much substance to this perspective. However, it is not a causal explanation of why the physical device meets its functional specification. It simply says there has to be some kind of causal connection, without saying in any detail what that might be. So how do we provide the details?

Consider our electronic case. Here the electronic circuit is a mirror image of the abstract digital one, and we are tasked with providing an explanation of why the electronic circuit satisfies the functional demands. Kroes [137, 139] provides an explanation for such technical artifacts in terms of physical laws, the physical makeup and configuration of the artifact, and the dynamic behaviors and causal interactions of its components. On his account, an explanation for why a designed digital circuit satisfies its functional specification would involve the following components:

- laws of physics and electronics;
- the physical make-up and configuration of the given electronic circuit;
- dynamic behaviors and causal interactions that govern the behavior of electronic circuits.

With the appropriate physical laws and structural configurations of the circuit, this yields the causal mechanisms that explain why the electronic device behaves the way it actually does. Such an account would go beyond the simple appeal to counterfactuals, and provide a more detailed causal explanation. In particular, the laws that govern digital circuits come to the fore; by appeal to these, and the actual digital structure, we are able to provide an explanation of what the electronic device will actually do.

Of course, as Kroes points out, this explains what it actually does, not necessarily what it was intended to do. It does not follow from an explanation along these lines that the function of our electronic device is to compute addition or logical conjunction. It is impossible to get the normative explanandum containing the ascription of a proper function from the purely descriptive explanans. Kroes concludes that the explanation as presented is not a technological explanation, since it does not properly account for function in terms of its structure. However, he argues that the relation between abstract structure and physical devices can be conceived in terms of pragmatic rules of action that are grounded in such causal considerations. For example, if one's goal is to compute an AND gate, then something like the following causal conditional holds:

- If an AND gate is used properly in appropriate circumstances, by appeal to the above causal account, it will compute the Boolean operation of conjunction.

We may then infer the following rule of action:

- In order to compute *and*, use an AND gate.

In this context of action, the AND gate is a means to an end, and acquires a function. On this analysis, the rule of action is formulated on the basis of a causal conditional that was derived from the circuit's structure. The explanation is not a deductive explanation. Rather, it connects structure and function on the basis of pragmatic rules of action based on causal relations. This does offer a reasonable account of explanation for simple electronic devices. But nothing like this is at play in the case of the mathematical notion of correctness.

28.2 Mathematical Explanation

How do we explain how a program satisfies the goal set by the specification? There are two aspects to this. One must be a direct explanation of why the symbolic program meets its specification, and the other involves the correctness of the implementation. Since the latter will reduce to mathematical and technological explanations, we concentrate on the correctness of the symbolic program relative to its specification.

In this case, what constitutes an explanation? In the ideal case it is a proof that explains why a program meets its specification. However, some proofs are more explanatory than others. Exhibiting a one-to-one correspondence is a form of proof, but not a particularly explanatory one. Mark Lang [143] employs several examples to illustrate the difference, but given the present application, the following is the most relevant.

In the following table, we can form a six-digit number by taking the three digits in any row, column, or main diagonal in forward and then reverse order:

7	8	9
4	5	6
1	2	3

As you can easily verify, every number so calculated is divisible by 37. But what is the explanation? A proof that checks each of these calculations separately does not explain why every calculator number is divisible by 37. Compare this case-by-case *proof* with the following.

The three digits from which a calculator number is formed are three integers a, $a+d$, and $a+2d$ in arithmetic progression. Take any number formed from three such integers in the manner of a calculator number — that is, any number of the form

$10^5 a + 10^4(a + d) + 10^3(a + 2d) + 10^2(a + 2d) + 10(a + d) + a$. Regrouping, this is equal to

$a(10^5 + 10^4 + 10^3 + 10^2 + 10 + 1) + d(10^4 + 2.10^3 + 2.10^2 + 10) = 111111a + 12210d = 1221(91a + 10d) = (3 * 11 * 37)(91a + 10d)$.

This proof explains why all the calculator numbers are divisible by 37. This proof, unlike the case-by-case one, reveals the result to be no coincidence. The brute force proof explains nothing, and this is similar to the situation with the square root example.

So what is an explanatory proof? Steiner [211] proposed an account in which he suggested that to explain the behavior of an entity one must locate its characterizing property i.e., a property unique to a given entity or structure within a family or domain of such entities or structures. Explanatory proofs involve or appeal to these characterizing properties. In Steiner's words:

> An explanatory proof makes reference to a characterizing property of an entity or structure mentioned in the theorem, such that from the proof it is evident that the result depends on the property.

One could develop an account of explanation for correctness proofs along these lines, but there is a rather different notion that is better suited. Lang [143] argues that in a certain kind of explanatory proof there is a symmetry between the statement of the theorem and the proof itself:

> Often a mathematical result that exhibits symmetry of a certain kind is explained by a proof showing how it follows from a similar symmetry in the problem. Each of these symmetries consists of some sort of invariance under a given transformation; the same transformation is involved in both symmetries.

Such symmetries consist of an invariance under a transformation where the same transformation is involved in the problem statement and the proof.

But what might this notion of symmetry mean for correctness proofs? Consider the Hoare calculus where the theorem statements take the following form.

$$\{\phi\}P\{\psi\}\,.$$

Recall that this asserts that if the predicate calculus assertion ϕ is true before the program P runs, then ψ will be true afterwards. The statement of the theorem thus takes the form of such an assertion. The Hoare calculus supplies rules for all constructs in the language of this form. For example, the rule for sequencing is given as follows.

$$\frac{\{\phi\}P\{\psi\} \qquad \{\psi\}Q\{\mu\}}{\{\phi\}P; Q\{\mu\}}\,.$$

So for sequencing, the output state for the first program becomes the input state for the second. This can be rephrased in explanatory terms as follows.

- In the state ϕ, $P; Q$ will end in state μ because, in state ϕ, the program P will end in state ψ, and in state ψ the program Q will end in state μ

This is a deductive explanation of why the program for sequencing meets its specification in transforming state ϕ into state μ. Each step in the formal proof relates directly back to the symmetry of the theorems statement. So, if we follow the whole proof through, we see that each step mirrors the statement of the theorem. There are other characteristics of explanatory proofs in mathematics, but the above seems appropriate in that the whole proof is structured around the symmetry induced throughout the proof by the Hoare triples. Each step is a miniature form of the theorems statement. There is a uniformity to the proof that centers on the statement of correctness, and the correctness of the whole is compositionally constructed in a way that follows the structure of the program. In this sense, the proof provides a transparent explanation of why the program satisfies its specification.

28.3 Structural Explanation

The third case concerns software systems. At the top level, we are required to explain why the system specification is satisfied by the system design. On the face of it, what seems most appropriate here is some form of structural explanation [87, 151, 186]. Piccinini [186] argues that computational explanations are always mechanistic in the sense of [20]. In [186] Piccinini defines a computing mechanism as a mechanism whose functional organization brings about computational processes. According to [151] a mechanism can be defined as a set of components whose functional capabilities and organization enable them to bring about an empirical phenomenon. Something like this seems to be at play with large software systems at the design stage.

Consider again the demands of our rudimentary library system:

A town requires a new library system. It must support the cataloging of new books, and the borrowing and returning of books by readers. It must have a system for acquiring books from vendors, and it must support all the normal security and financial arrangements concerning the borrowing and returning of books.

Now consider our UML design shown in Fig. 6.1. What kind of explanation is relevant here? How do we argue that this design meets the functional demands? Here the design almost wears its explanation on its sleeve. The various classes and their structural relationships would seem to explain how readers borrow and return books. There is a kind of structural explanation that the various demands are met by the classes and their interrelationships.

Of course, these structural connections only take us so far. They are a top-level form of explanation. The actual verification of software employs a host of techniques, and there are hidden forms of explanation in software verification that involve abstraction and modeling. Explaining the behavior of large complex systems often requires one to construct a formal model of the system that abstracts away from the noise and details of the underlying implementation. But these relate to the next level of implementation where the UML specifications are implemented in code. Angius and Tamburrini [10, 11], provide a detailed analysis of these forms of explanation.

28.4 Conclusion

There are many other avenues that we would need to explore for a more complete account of computational explanation. Indeed, the topic of computational explanation is large, and this chapter pretends to be nothing more than a taster. Our main focus here is to explain away the simple mapping puzzle both at the mathematical and at the physical levels. But there is still a missing ingredient in our account of correctness; we seem to have lost sight of the intentional aspect.

Chapter 29
INTENTION AND CORRECTNESS

The simple mapping account of correctness reduces correctness to extensional agreement. While we have indicated various means of cutting down the number of solutions, restricting them in some way, there is something more substantial missing from our account. An assumption underlying the simple mapping account is that computation and programming are notions that can be fully characterized independently of any intentions. Wittgenstein implicitly criticized this long before the simple mapping account came on the scene:

> There might be a caveman who produced regular sequences of marks for himself. He amused himself, e.g., by drawing on the wall of the cave. But he is not following the general expression of a rule. And when we say that he acts in a regular way that is not because we can form such an expression. That is, the fact that we could construct a rule to describe the regularity of his behaviour does not entail that he was following that rule [257].

Imagine a caveman accidentally scratches the following table in the sand with a stick:

1	2	3	4	5
1	1.41421356237	1.73205080757	2	2.2360679775

Do we say he has correctly programmed some fragment of the square root? Obviously not: he had no intention of constructing a program for the square root. Just getting the answer right, by accident, involves no intention of any kind. The caveman is merely scribbling in the sand. Programming and computing are intentional activities, and establishing that a program meets its specification is part of the activity of programming. Hiving off the extensional aspect of this activity, and considering it in isolation from its intentional context, is at the core of the problem raised by the simple mapping account. Removing agency and intention is to treat the program as a thing with no function. But programs as technical artifacts are things with function, and the latter is intentional in nature. A program or computation is a technical artifact with a function as well as a structure, and a function

© Springer-Verlag GmbH Germany, part of Springer Nature 2018
R. Turner, *Computational Artifacts*, https://doi.org/10.1007/978-3-662-55565-1_29

is not just an expression that articulates the conditions of correctness; it is the intention of some agent. But now we require an account of intention; we require an intentional theory of function.

29.1 The Standard Account

On the intentional theory of function, it is an agent that ascribes functions to artifacts: functions reflect the intentions of a designer. A toothbrush can serve as a cutlery cleaner or a tooth cleaner. The function of an artifact is dependent upon the decision of some agent – the one who decides it will be a toothbrush or a cutlery cleaner:

> [t]he function of an artifact is derivative from the purpose of some agent in making or appropriating the object; it is conferred on the object by the desires and beliefs of an agent. No agent, no purpose, no function [152].

Without such intentions there is no function. Unfortunately, the introduction of agents and intentions poses a significance concern. In their crude form, such theories have difficulty in accounting for how they impose any constraints upon the actual thing that is the artifact. How do they guarantee extensional agreement between specification and structure?

> If functions are seen primarily as patterns of mental states, and exist, so to speak, in the heads of the designers and users of artifacts only, then it becomes somewhat mysterious how a function relates to the physical substrate in a particular artifact [138].

> The question is: under what circumstances could we say that he was following a rule? Let us consider very simple rules. Let the expression be a figure, say this one: |--| and one follows the rule by drawing a straight sequence of such figures (perhaps as an ornament). Under what circumstances should we say: someone gives a rule by writing down such a figure? Under what circumstances: someone is following this rule when he draws that sequence? It is difficult to describe this. The crucial point in what follows is that the answer to this question has nothing whatsoever to do with any putative central events which might accompany such regular behaviour. If one of a pair of chimpanzees once scratched the figure |--| in the earth and thereupon the other the series |--| |--| etc., the first would not have given a rule nor would the other be following it, whatever else went on at the same time in the mind of the two of them [255].

Intentional theories have to account for how the connection with the physical structure is determined. Unfortunately, as an agent, I can have all sorts of images in my head. How do any such images connect with the structure of an artifact? How do they deliver extensional agreement? While the latter may not be sufficient to provide an adequate account of correctness, it is certainly necessary.

29.2 Definition and Intention

In our account of specification, there are two components:

- definition;
- intention.

The former contains the propositional content of the specification, and the latter the intentional stance of the agent. The intention gives governance of the definition over the artifact, and thus provides its correctness conditions. Extensional agreement is delivered by the definition that forms the propositional content of the specification. The intention is limited to the act of giving the definition governance; it is constituted by the intentional stance of the agent. This is inspired by the account of intention given in [182].

Why does this help? The propositional content of the definition is fixed by its semantic interpretation, and this is given independently of the agent. Consequently, the mental states of the agent can have no influence on the determination of this content. Whatever the semantic account of the language of the definition, it is not to be given by the mental states of any agent. It is a normative account that provides criteria of correct use.

But what if the agent misunderstands the propositional content of the specification? Indeed, is the very demand for such a normative account too demanding?

Chapter 30
RULE FOLLOWING AND CORRECTNESS

Programming involves the construction of programs that meet their specifications. This draws in all the associated activities of programming such as testing, verification, and establishing correctness. And these are rule following activities.

30.1 Computing and Programming

Ordinary people and, in particular, mathematicians calculated long before anyone wrote computer programs. For example, consider the standard algorithm for multiplication:

1. Multiply the multiplicand by the least significant digit of the multiplier to produce a partial product.
2. Continue this process for all higher order digits in the multiplier.
3. Right align each partial product with the corresponding digit in the multiplier.
4. Sum the partial products.

Such algorithms lay out rules in a systematic way, and to compute with the algorithm is to follow the rules. Programming takes a further step: *we get a machine to do the computing for us.* But programming still involves rule following. The programmer is constrained by her programming language, and the specification of the required program. To construct a program, the programmer must obey the rules laid down in the syntax of the host programming language. The grammar of a programming language is a set of rules that determine what constitutes a well-formed program. The syntax analyzer of the implementation will check that these rules have been followed. Furthermore, in order to construct a program that meets the given specification, the programmer must follow the semantic rules of the programming language: she must use the constructs in accord with the semantics to ensure that the program does what is required by the specification. In doing, so she has also to follow the semantic rules of the specification language.

© Springer-Verlag GmbH Germany, part of Springer Nature 2018

R. Turner, *Computational Artifacts*, https://doi.org/10.1007/978-3-662-55565-1_30

So, whatever level of formality is adopted for the semantic description, vernacular or formal, programming involves a great deal of rule following.

30.2 Rule Following

In Wittgenstein's *Philosophical Investigations*, a student is taught how to continue a series of natural numbers in accordance with the rule *add2*. He applies the rule, and to begin with, he gets it right: $0, 2, 4, 6, 8, 10, 12, \ldots, 1000$. But from 1000 onward, he continues with $1004, 1008, 1012, \ldots$. The teacher informs him of his mistake, but the student insists that he is following the rule. However, from his answers, he appears to be using the following rule: add 2 for each step up to 1000, then add 4 up to 2000, after that add 6 up to 3000, and so on. Such an interpretation of the rule matches all the instances he has been shown by the teacher. Seemingly, every rule, no matter how simple, may be read in a nonstandard way:

> This was our paradox: no course of action could be determined by a rule, because every course of action can be made out to accord with the rule. The answer was: if everything can be made out to accord with the rule, then it can also be made out to conflict with it. And so there would be neither accord nor conflict here [255].

Apparently, any interpretation can be reinterpreted in a different way without failing to match all the previous applications of the rule:

> However many rules you give me – I give a rule which justifies my employment of your rules [257].

The expression of a rule does not dictate the whole range of correct instances in all future circumstances. In teaching the *add2* rule, it is not possible to explain all the steps in advance:

> How was it possible for the rule to have been given an interpretation during instruction, an interpretation which reaches as far as any arbitrary step [257]?

Seemingly, interpretations cannot fix matters.

How does this impact upon our theory of function that restricts the intentional component of function to the intentional stance of the agent? Presumably, we may conclude from our rogue calculator that our strategy will not work. Whatever the actual propositional content of the specification, the agent may well be interpreting matters in a completely different way. He may think that any program which meets his current interpretation of the rule *add2*, is correct. Extensional agreement is still lost.

30.3 The Skeptical Paradox

But, according to Kripke's reading of Wittgenstein, matters are much worse. Not only is our agent free to choose any interpretation, but also there is no fact of the matter about which is the right one. Kripke gives a slightly different example to illustrate the reasoning that leads to what has become known as the *skeptical paradox*. Suppose that you have never added numbers greater than 50 before, and you are asked to add 68 and 57. You do so, and get the correct answer, 125. But now a bizarre friend argues:

1. There is no fact about your past usage of the addition function that fixes the correct answer as, 125.
2. Indeed, nothing justifies you in giving this answer rather than another.

You have never added numbers greater than 50 before, so it is consistent with your previous use of *plus* that you actually meant it to mean the *quus* function.

$$quus(x, y) = \begin{cases} x + y & if\ x, y < 57 \\ 5 & otherwise \end{cases}$$

Kripke argues that following a rule correctly is not justified by any fact that obtains between an agents particular application of a rule and the rule itself:

> Given that everything in my mental history is compatible both with the conclusion that I meant plus and with the conclusion that I meant quus, it is clear that the skeptical challenge is not really an epistemological one. It purports to show that nothing in my mental history of past behavior – not even what an omniscient God would know – could establish whether I meant plus or quus. But then it appears to follow that there was no fact about me that constituted my having meant plus rather than quus [134].

> Wittgenstein holds, with the skeptic, that there is no fact as to whether I mean plus or quus [134].

One might object that the addition function is not defined by example, but by an algorithm or function that inherits its meaning from the expressions of the language in which it is expressed. But then the language itself, its terms and expressions, are susceptible to different and incompatible interpretations, so the problem resurfaces at a higher level. Rules for interpreting rules can be interpreted in different ways.

Unfortunately, Kripke's interpretation of Wittgenstein's rule following has devastating consequences: it implies that there is no such thing as meaning something by an expression or sentence. For Kripke, normativity of meaning depends upon there being a correct interpretation. This has impact upon our account of the intentional theory of function. We have taken meaning to be normative: it must provide correctness conditions. If there cannot be rules governing the use of language, as the rule following paradox apparently shows, this is undermined.

30.4 The Skeptical Solution

Kripke embraces the paradox, and offers a corresponding solution that is not based on some mental state of meaning, interpretation, or intention. It is a skeptical solution to a skeptical problem. He argues that the assertion that the rule is being followed is justified relative to the behavior of the agent. For a specific rule following incident, such behavior must be consistent with the expectations of other language users. Ascriptions of meaning and rule following have a non-fact-stating function: the use of the *plus* sign must agree with that of our arithmetical community. Consistent with this, Kripke argues that the notions of rule following and meaning only make sense relative to a community of rule-followers or speakers.

But does this save our practices? Does this provide us with a normative account with conditions of correct use? Miller [165] argues that it does not. He argues that statements with non descriptive semantic functions must also be assessed as either correct or incorrect. This suggests that a non factualist response to Kripke's skeptical paradox about meaning will not be sufficient to avoid the conclusion that all language is meaningless. Non-fact-stating language is rule-governed, and hence susceptible to the argument of the rule following skeptic. The skeptical argument undermines the general notion of a rule, not just the notion of a rule governing the use of expressions with descriptive semantic functions. If this is right, the strategy of the skeptical solution is ultimately self-defeating.

30.5 Rules and Customs

In contrast, John McDowell argues that Wittgenstein rejects the paradox, and offers a *straight* solution. The paradox is rejected because it assumes that in order to understand something, we must have an interpretation of it. And it is this assumption that is false:

> That is, whatever piece of mental furniture I cite, acquired by me as a result of my training in arithmetic, it is open to the sceptic to point out that my present performance keeps faith with it only on one interpretation of it, and other interpretations are possible. So it cannot constitute my understanding 'plus' in such a way as to dictate the answer I give. Such a state of understanding would require not just the original item but also my having put the right interpretation on it. But what could constitute my having put the right interpretation on some mental item? And now the argument can evidently be repeated [161].

So, for McDowell the central issue is to show how one can understand something without *interpreting* it. If we ditch interpretation as the mechanism of understanding, what are we left with? McDowell argues that to understand rule following we should understand it as resulting from inculcation into a custom or practice. To understand addition is simply to have been inculcated into a practice of adding. McDowell's Wittgenstein invites us to see the significance of rules within the practices in which the rule is normally and naturally applied. McDowell's Wittgenstein invites us to see the significance of rules not in a detached manner, but in the customary practices in which the rule's applications and functions are salient. Instead of conceiving practice as merely some manifestation of grasping a rule, whether one really understands the rule depends on whether one responds in the same way as we are used to doing in various contexts in ordinary life. Grasping a rule is

> exhibited in what we call "obeying the rule" and "going against it" in actual cases [255].

Programming as a rule based activity is manifested in the associated behavior of the programmer.

> If however there were observed, e.g., the phenomenon of a kind of instruction, of shewing how and of imitation, of lucky and misfiring attempts, or reward and punishment and the like; if at length the one who had been so trained put figures which he had never seen before one after another in sequence as in the first example, then we should probably say that the one chimpanzee was writing rules down, and the other was following them [255].

We cannot do justice to the subtlety and depth of McDowell's account contained in [157, 158, 159, 160, 161]. However, the central point seems clear: the paradox evaporates if we can find an account of understanding that does not assume that such understanding is provided by interpretation.

30.6 Rule Following and Computer Science

This debate is central to any serious philosophy of computer science. Programming is a rule governed activity, and involves the rules embodied in the syntax and semantics of a programming language. Without a normative semantic theory for programming languages, there can be no notion of correctness for programs.

Moreover, the implications affect the central concerns of the philosophy of technology. The standard worry about intentional theories of function is not centrally a worry about mental states, and how they are able to reflect the structure of an artifact. It is a worry over the very meaning of the language that expresses functional demands. It is a concern over how any possible articulation of the function of an artifact can have normative status.

This is a complex and ongoing debate. Our objectives here are modest: we have tried to put enough bones on matters to indicate how it affects the issue of program correctness.

However, it does seem fitting that we end with an issue at the heart of the contemporary philosophies of language, mind, logic, mathematics, technology, and computer science.

Nomenclature

Architectural description languages. Language designed for the specification of software architectures.

Classes and objects. These are the data types of objected-oriented programming languages. They package all the information about a class of objects (e.g., cars) in terms of their attributes and performable operations. See Chapter 8.

Functional paradigm. The family of programming languages where the core notion is that of a mathematical function. See Chapter 8 for further details.

Hardware description language. These languages are introduced in Chapter 4. However, a good introduction can be found at

http://user.engineering.uiowa.edu/vlsi1/notes/Verilog.pdf

High-level programming language. A language that is at a conceptual distance from the physical machine. These languages involve abstract notions that have no direct interpretation on physical devices. See Chapter 8 for more explanation.

Implementation. Implementation is used for a variety of processes in computer science. But perhaps the most common use concerns high-level languages and how they are interpreted on an actual machine. See the chapter on implementation and semantics for more details.

Intuitionistic logic. The difference between classical and intutionistic logic concerns the law of excluded middle, i.e., every proposition or its negation is provable.

Levels of abstraction. Modern software is built upon translations between languages that operate with different control constructs and different methods of data representation. High levels of abstraction involve more abstract notions. See Chapter 21 for more explanation.

Object-oriented programming. Object oriented programming (OOP). Languages for OOP include Java, C++, and Python.

Parameter mechanism. The order or means of evaluating the arguments to a function or procedure.

© Springer-Verlag GmbH Germany, part of Springer Nature 2018

R. Turner, *Computational Artifacts*, https://doi.org/10.1007/978-3-662-55565-1

Polymorphism. Polymorphic functions operate on families of types, members of which are presented as arguments to the function. For example, they might operate on numbers, lists, strings, etc.

Precondition. Preconditions are the conditions that, together with its type, determine the scope of a program. We shall introduce this formally in chapter 12.

Procedure. Procedures form one of the earliest abstraction mechanisms in programming languages. They are packaged chunks of code that can be named and later employed. More information can be found in Chapter 8.

Race condition. Processes competing for resources are said to race each other.

Rules of type inference. These are the rules that determine the type of objects in the language. Pierce's text [172] provides a good general introduction.

Rules of type inference. These are the rules that determine the type of objects in the language. Pierce's text [187] provides a good general introduction.

Security. Security is linked to the ability of a language to avoid errors. Much of this has to do with its type-checking facilities. See Chapter 26.

Software system. The system of software components that forms part of a whole computer system.

Specification. A specification is intended to say what a program is intended to do. See Chapter 14 for more explanation.

Turing complete. A programming language is said to be Turing complete if it can faithfully represent any given Turing machine.

Type inference. Type systems add some semantic content to the grammar of a programming language. They ensure that programs obey the type discipline of the language, and rule out some conceptual nonsense such as trying to add the Boolean value true to 4. See Chapter 26.

References

1. Abelson, H., Sussman, G.J., Sussman, J. *Structure and Interpretation of Computer Programs*. MIT Press (1985). 1.1

2. Abramsky, S., Jagadeesan, R., Malacaria, P. Full abstraction for PCF. In *Theoretical Aspects of Computer Software*. M. Hagiya, J. C. Mitchell (eds.). Elsevier (2000). 9, 10.5

3. Abramsky, S., Jung, A. Domain Theory. In *Handbook of Logic in Computer Science*, Volume 3. S. Abramsky., D. Gabbay., T. Maibaum (eds). Clarendon Press, Oxford (1994). 14.4, 19.4

4. Abramsky, S., McCusker, G. Games and Full Abstraction for the Lazy Lambda Calculus. In *Proceedings of the Tenth Annual Symposium on Logic in Computer Science* (1995). pp. 234-243. IEEE Computer Society Press. 9, 10.5

5. Abramsky, S., Jagadeesan, R. Games and Full completeness for multiplicative linear logic. In *Foundations of software technology and theoretical computer science* (1992). pp. 291-301. 10.5

6. Abrial, J. *The B-Book*. Cambridge University Press (1996). 7.1

7. Agarwal, A. *Foundations of Analog and Digital Electronic Circuits*. The Morgan Kaufmann Series in Computer Architecture and Design (2005). 4.3, 27.1

8. Aho, A., Lam, R., Sethi, R., Ullman, J. D. *Compilers: Principles, Techniques, and Tools*. Pearson (2013). 11

9. Allen, R. A. Formal Approach to Software Architecture. Robert J. Allen. Ph.D. Thesis, Carnegie Mellon University (1997). CMU Technical Report. 7.1, 15

10. Angius, N., Tamburrini, G. Scientific theories of computational systems in model checking. Minds and Machines 21 (2), 323-336 (2011). 25.4, 28.3

11. Angius, N., Tamburrini, G. Explaining Engineered Computing Systems. Behaviour: the Role of Abstraction and Idealization. Philosophy & Technology, 1-20 (2016). 28.3

12. Arkoudas. K., Bringsjord, S. Computers, Justification, and Mathematical Knowledge. Minds & Machines, 17:185-202 (2007). 25.2

13. Ashenden, P. *The Designer's Guide to VHDL*. The Morgan Kaufmann Series in Computer Architecture and Design (2008). 4.2, 7.1

14. Backus, J. Algol 60 Report. Communications of the ACM CACM, Volume 3 Issue 5. pp 299-314 (1960). 8.1

15. Baier, C., Katoen, J.P. *Principles of Model Checking*. MIT Press (2008). 25.2

16. Baljon, C. J. History of History and Canons of Design. Design Studies, 23(3), 333-343 (2002). 15

17. Baker, A. Simplicity. *The Stanford Encyclopedia of Philosophy* (Winter 2016 Edition). 2.3, 16

18. Barendregt, H. P. Lambda Calculi with Types. In *Handbook of Logic in Computer Science*. Vol. III. Ed. S. Abramsky, D. Gabbay, T. Maibaum. Oxford Univ. Press (1992). 8.2, 20.3

19. Barker-Plummer, D., Barwise, J., Etchemendy, J. *Language, Proof and Logic*. CSLI publications (1999). 12.1

20. Bechtel, W., Abrahamsen, A. Explanation: A mechanist alternative. Studies in History and Philosophy of Biological and Biomedical Sciences, 36. 421-441 (2005). 28.3

21. Bentley, J., Doug McIroy. Engineering a Sort Function. Software Practice and Experience, vol. 23(11), 1249-1265 (1993). 16.1

22. Bloch, J., and Venners, B. Bloch on Design: A Conversation with Effective Java Author, Josh Bloch by Bill Venners. JavaWorld, January 4 (2002). 26.3

23. Boehm, H., Demers, A., Donahue, J. An Informal Description of Russell. Tech. Rep. 80-430, Comp. Sci. Dept., Cornell University (1980). 20.1

24. Boghossian, P. The Rule-following Considerations. Mind, 507-549 (1989). 9.2

25. Booch, G., (2007). *Object-Oriented Analysis and Design with Applications*. Addison-Wesley (2007). 17, 17.1

26. Börger, E., Schulte, W. A Programmer Friendly Modular Definition of the Semantics of Java. In *Lecture Notes in Computer Science* (2007). Vol. 1523. Pages: 353-404. 10

27. Britannica. Computer Programming Language. `http://www.britannica.com/technology/computer-programming-language`. Last accessed December 31, 2017. 8

28. Buckminster Fuller, R. On Beauty. `https://www.brainyquote.com/authors/r_buckminster_fuller`. Last accessed December 31, 2017. 16.1

29. Burgess, J. P. *Fixing Frege*. Princeton University Press (2005). 21.3

30. Burgess, J. and Rosen, G. *A subject With No Object*. Oxford University Press (1997). 21.1

31. Burge. T. Computer Proofs, A priori Knowledge, and Other Minds. Phil. Perspectives 12 (1998). 25.2

32. Camurati, P., Prinetto, P. Formal verification of hardware correctness: introduction and survey of current research. Computer (1988). 21(7):8-19. 27

33. Cardelli, L., Abadi, M. *A Theory of Objects*. Monographs in Computer Science. Springer (1996). 7.1

34. Carruthers, P. The case for massively modular models of mind. In Robert J. Stainton (ed.), *Contemporary Debates in Cognitive Science*. Blackwell (2006). 2.3, 17.5

35. Chalmers D. (1996). Does a Rock Implement Every Finite-State Automaton? Synthese (2): 345-363 (1996). 28.1

36. Church, A. An unsolvable problem of elementary number theory. American Journal of Mathematics (1936). 58108:309-33. 8.2, 10.4

37. Clark, R. G., Moreira, A. Formal Specifications of User Requirements. Automated Software Engineering (1999). Volume 6, Issue 3. 14.1

38. Clarke, P., Gladstone, C., MacLean, A. C. N. SKI – The S, K, I Reduction Machine. LISP Conference, pages 128-135 (1980). 11.1

39. Coad, P. and Yourdon, E. *Object Oriented Analysis*. Yourdon Press (1990). 14.5

40. Cobham, A. The intrinsic computational difficulty of functions. *Proceedings of the 1964 International Congress*, pp. 24-30. (ed.) Y. Bar-Hillel. Logic, methodology and philosophy of science, North-Holland (1964). 23.2

41. Colburn, T., Shute, G. Abstraction in Computer Science. Minds & Machines, 17:169-184 (2007). 2.3

42. Colburn, T. *Philosophy and Computer Science*. M.E. Sharp Publishers (2000). New York, London. 5, 5.4, 11.3

43. Colburn, T., Shute, G. Metaphor in Computer Science. Journal of Applied Logic, Volume 6, Issue 4, pp 526-533 (2008). 9.4

44. Constantine, L. and Yourdon, E. *Structured Design: Fundamentals of a Discipline of Computer Program and Systems Design* (1979). 17.6

45. Copeland, J. What is computation? Synthese, Volume 108, issue 3, pp 335-359 (1996). 27.2, 28.1

46. Cross, N. A history of design methodology. In *Design methodology and relationships with science*. Kluwer Academic Publishers, Dordrecht (1993). IV

47. Cummins, R. Functional Analysis. Journal of Philosophy, 741-765 (1975). 3.3

48. The WHILE programming language. *A Dictionary of Computing.* Oxford University Press (2004). 5.2, 8.1

49. Darwin: an architecture description language. `https://en.wikipedia.org/wiki/Darwin_(programming_language)`. Last accessed December 31, 2017. 1.3, 7.1

50. Davis, M. *The Undecidable: Basic Papers on Undecidable Propositions, Unsolvable Problems, and Computable Functions.* Dover (1965). V

51. De Bakker, J.W. (1980). *Mathematical Theory of Program Correctness.* Prentice-Hall (1980). 9

52. De Millo. R.A., Lipton, R.J., Perlis, A. J. Social Processes and Proofs of Theorems and Programs. Communications of the ACM CACM. Volume 22 Issue 5. pp 271-280 (1979). 25.1

53. DeMarco, T. *Structured Analysis and System Specification.* Prentice Hall (1979). 17, 17.5

54. Dijkstra, E.W. The Humble Programmer. Turing Award Lecture. Printed in *Classics in Software Engineering.* Yourdon Press (1979). 14.5, 16, 18.8

55. Dijkstra, E.W. *Structured programming.* Academic Press Ltd. (1972). 17

56. Dijkstra, E.W. On a Cultural Gap. The Mathematical Intelligencer 8. 1: 48-52 (1986). 1.1

57. Dijkstra, E.W. Quotes. `https://en.wikiquote.org/wiki/Edsger_W._Dijkstra`.Last accessed December 31, 2017. 1.2, 2.3

58. Dijkstra, E.W. The Next Fifty Years. EWD1243 `https://www.cs.utexas.edu/users/EWD/transcriptions/`. Last accessed December 31, 2017. 16

59. Dijkstra, E.W. Programming as a discipline of mathematical nature. American Mathematical Monthly, Vol. 81. No. 6, pp. 608-612 (1974). 18.8

60. Dijkstra, E.W. On the Cruelty of Really Teaching Computer Science. `https://www.cs.utexas.edu/~EWD/ewd10xx/EWD1036.PDF` 18.8

61. Dijkstra, E.W. Quotes. `https://www.goodreads.com/author/quotes/1013817`. Last accessed December 31, 2017. 1.1, 25.3

62. Duhem, P. (1954). *The Aim and Structure of Physical Theory.* Princeton University Press (1991). 11.3

63. Edmonds, J. (1965). Paths, trees, and flowers. Canad. J. Math. 17, 449-467 (1965). 23.2

64. Emde Boas, P. van. Machine models and simulations. In J. Van Leeuwen (ed.), *Handbook of theoretical computer science* (vol. A): algorithms and complexity, pp. 1-66. MIT Press (1990). 23.2

65. Encapsulation (computer programming). https://en.wikipedia.org/wiki/Encapsulation (computer programming) 17.1

66. Meyer, B. *Eiffel: The Language.* Prentice Hall object-oriented series (1992). 8.4

67. Felleisen, M. On the Expressive Power of Programming Languages. In *proceedings of the third European symposium on programming on ESOP '90*, pp 134-151. Springer (1990). 19.5

68. Fernandez, M. *Programming Languages and Operational Semantics: An Introduction.* King's College Publications (2004). 2.1, 9, 10

69. Fernández, Maribel. *Models of Computation. An Introduction to Computability Theory.* Springer (2009). 8

70. Fetzer. J.H. Program Verification: the very idea. Communications of the ACM. Volume 3:9 (1988). 2.4, 11.3, 24

71. Floyd, R.W. Assigning meanings to programs. *Proceedings of the American Mathematical Society Symposia on Applied Mathematics*, Vol. 19, pp. 19-31 (1967). 15.3

72. Floridi, L. Levellism and the Method of Abstraction. IEG – Research Report 22.11.04 (2004). 14.5

73. Fodor, J. A. *The Modularity of Mind: An Essay on Faculty Psychology.* MIT Press (1983). 2.3, 17.4, 17.5

74. Fowler, M. *UML Distilled: A Brief Guide to the Standard Object Modeling Language.* Addison-Wesley (2003). 7.1

75. Franssen, M., Lokhorst, G., van de Poel, J. Philosophy of Technology. *The Stanford Encyclopedia of Philosophy* (Fall 2015 Edition). 1, 2.2, II

76. Frege, G. *Posthumous Writings.* trans. by: Long and R. White. Oxford: Basil Blackwell (2003). 2.3, 9.3, 10.2

77. Fresco, N., Primiero, G. Miscomputation. Philosophy & Technology, 26, 253-272 (2013). 24.3

78. Frigg, R. Hartmann, S. and Imbert, C. Preface: special issue: models and simulations 2. Synthese, 180 (1) (2011). 25.4

79. Gabbrielli, M. *Programming Languages: Principles and Paradigms.* Springer (2010). 8

80. Gallier, J. *Logic for Computer Science: Foundations of Automatic Theorem Proving.* Dover (1985). 8.3, 8.3

81. Gamma, E., Helm, H., Johnson, R., Vlissides, J. *Design Patterns – Elements of Reusable Object-Oriented Software.* Addison-Wesley (1995). 17, 17.1, 20.3

82. Ganascia, J. G. Abstraction of levels of abstraction. Journal of Experimental & Theoretical Artificial Intelligence, 27(1):1-13 (2014). 14.5

83. Garlan, M. Acme: Architectural description of component-based systems. In *Foundations of Component-Based Systems*, chapter 3. Cambridge University Press (2000). 7.1

84. Garlan, D., Shaw, M. An Introduction to Software Architecture. CMU-CS-94-166 (1994).

85. Gelernter, D. *Machine Beauty.* Basic Books (1998). 16.1

86. Girard, J. Y. The system F of variable types, fifteen years later. Theoretical Computer Science, 45:159-192 (1986). 10.5

87. Glennan, S. Mechanisms and the nature of causation. Erkenntnis, 44(1) (1996). 28.3

88. Glüer, K., and Wikforss, Å., The Normativity of Meaning and Content. *The Stanford Encyclopedia of Philosophy* (2006). 9.2

89. Godfrey-Smith, P. Abstractions, Idealizations, and Evolutionary Biology. In *Mapping the Future of Biology: Evolving Concepts and Theories*, (pp. 47-56). A. Barberousse., M. Morange., T. Pradeu (eds.). Springer (2009). 27.2

90. Goguen, J. A., Burstall, R. M. A study in the foundations of programming methodology: specifications, institutions, charters and parchments. *Proc. Summer Workshop on Category Theory and Computer Programming.* Springer Lecture Notes in Computer Science, Vol. 240, 313-333 (1986). 21

91. Goldreich, O. *Computational Complexity: A Conceptual Perspective.* Cambridge University Press (2008). 1.2, 1.3

92. Gordon, M. J. C. *The Denotational Description of Programming Languages.* Springer (1979). 9, 10, 19.3

93. Gouthier, R. and Pont, S. *Designing Systems Programs* in C. Prentice-Hall (1970). 15

94. Graham, P. *On Lisp: Advanced Techniques for Common Lisp.* Perfect Paperback (1993). 8.2

95. Griffiths, P. Philosophy of Biology. *The Stanford Encyclopedia of Philosophy* (2017). 1

96. Grothendieck, A. *Récoltes et semailles: Réflexions et témoignage sur un passé de mathématicien* (2008). https://uberty.org/wp-content/uploads/2015/12/Grothendeick-RetS.pdf. Last accessed December 31, 2017. 25.1

97. Gunter, C.A. *Semantics of Programming Languages: Structures and Techniques.* MIT Press (1992). 10, 19.3

98. Gupta, A. Definitions. *The Stanford Encyclopedia of Philosophy* (2008). 5.1, 14.2

99. Guttag, J. V. *Introduction to Computation and Programming Using Python.* MIT Press (2013). 21.4

100. Hacking, I. *Why is there a Philosophy of Mathematics at all?* Cambridge University Press (2014). 25.1

101. Hale, B and Wright, C. The Metaontology of Abstraction. In *Metametaphysics: New Essays on the Foundations of Ontology.* pp. 178-212. D. J. Chalmers, D. Manley, R. Wasserman (eds). Oxford (2009). 21.2

102. Hamming, R.W. One Man's View of Computer Science. ACM Turing Lecture 1968. In *Turing award lectures,* ACM, New York (2007). 1.1

103. Hancock, J. Proceedings of National Science Foundation workshop 1986. Engineering Education: NSB 86-100 (1986). IV

104. Hankin, C. *An Introduction to Lambda Calculi for Computer Scientists*. King's College Publications (2004). 7.1, 8.2

105. Hanson, M. Rischel, H. *Introduction to Programming using SML*. Pearson (1999). 8.2

106. Hardy, G. H. *A Mathematician's Apology*. Cambridge (1946). 25.1

107. Hartmanis, J., Stearns, R. E. On the computational complexity of algorithms. Transactions of the American Mathematical Society, 117: 285-306 (1965). 23.1

108. Kevlin H. For the sake of simplicity (1999). https://www.artima.com/weblogs/viewpost.jsp?thread=362327. Last accessed December 31, 2017. 16.2

109. Henson, M. *Elements of Functional Languages*. Blackwell Scientific (1987). 18.5

110. Henson, M. Formal Specification and Development in Z and B. 2nd International Conference on Z and B. *Springer Monographs in Computer Science* (2002). 18.5

111. Hilbert, D. *The Foundations of Geometry*. 2nd ed. Chicago: Open Court (1980). 10.2

112. Highsmith, J. The Agile Manifesto. http://agilemanifesto.org. Last accessed December 31, 2017. 6.1, 13.3, 13.4

113. Hill, R. Elegance in Software. Third Conference on the History and Philosophy of Programming. Paris, 25 June 2016. http://hapoc.org/sites/default/files/Hill_Elegance_in_Software_HaPoP3.pdf. Last accessed December 31, 2017. 16.1

114. Hoare, C.A. An Axiomatic Basis for computer programming. Communications of the ACM, Volume 12, Issue 10 (1969). 9, 10, 18.4, 24

115. Hoare, C. A. R., (1986). *Mathematics of Programming*. Prentice Hall (1986). 10.6, 18, 25.3

116. Hoare, C. A. R., (1980). Turing Award Lecture. Communications of the ACM 24 (2) (1981): pp. 75-83. 19.1, 19.4

117. Hoare, C. A. R. http://www.azquotes.com/quote/680860. Last accessed December 31, 2017. 1.3, 2.3, 15.2, 16, 21.1

118. Hoare, C. A. R. *Communicating Sequential Processes*. Prentice Hall International (2004). 7.1, 8.5

119. Hofmann, A. Quotes. https://en.wikiquote.org/wiki/Hans_Hofmann. Last accessed December 31, 2017. 16.2

120. Horning, J. http://www.azquotes.com/author/42137-Jim_Horning. Last accessed December 31, 2017. 12, 18.7

121. Horsten, L. Philosophy of Mathematics. *The Stanford Encyclopedia of Philosophy* (Winter 2016 Edition). 1

122. Hyland, J. and Ong, C. On Full Abstraction for PCF: I, II, and III. Information and computation, 163 (2), 285-408 (2000). 10.5

123. Indurkya, B. Metaphor and Cognition. Studies in Cognitive Systems 13. Springer (1975). 9.4

124. Indurkhya, B. Some philosophical observations on the nature of software and their implications for requirement engineering. https://www.academia.edu/7817075/. Last accessed December 31, 2017. 5, 5.4

125. Irmak, N. Software is an Abstract Artifact. Grazer Philosophische Studien, 86(1):55-72 (2012). 5, 5.4

126. Iverson, K. Turing Award lecture, Communications of the ACM, 23 (8), pp. 444-465 (1980). 25.4

127. Jones, C. B. *Systematic Software Development Using VDM*. Prentice Hall International (1990). 1.3, 2.3, 5.1, 7.1, 12, 12.2, 12.5

128. Jones, C. Software Estimating Rules of Thump. Computer, Volume: 29, Issue: 3 (1996). 24

129. Junger, P. Dual Nature of Computer Programs. http://samsara-blog.blogspot.co.uk/2006/01/dual-nature-of-computer-programs-redux.html. Last accessed December 31, 2017. 5

130. Kernighan, B. and Ritchie, D. *The C Programming Language*. Prentice Hall (1988). 19.2

131. Kleene, S. C. Origins of recursive function theory, Ann. Hist. Comput. 3, 52-67 (1981). 22.3

132. Knuth, D. Quotes on Computer Programming. http://www.azquotes.com/author/8177-Donald_Knuth. Last accessed December 31, 2017. 1.2

133. Kochan, S. *Programming in C*. Developer's Library (2004). 8.1

134. Kripke, S. *Wittgenstein on Rules and Private Language*. Harvard University Press (1982). 11.3, 30.3

135. Kroes, P. and Meijers, A. The Dual Nature of Technical Artifacts – presentation of a new research programme. Techné: Research in Philosophy and Technology, Vol 6 (2002). 3.1

136. Kroes, P., and Meijers, A. The dual nature of technical artifacts, Stud. Hist. Phil. Sci. 37:1–4 (2006). 3.1, 3.3

137. Kroes, P. Technological explanations: the relation between structure and function of technological objects. Techné: Research in Philosophy and Technology 3 (3), 124-134 (1988). 28.1

138. Kroes, P. Engineering and the Dual Nature of Technical Artefacts. Camb. J. Econ., 51-62 (2010). 3.1, 3.3, IV, 15, 29.1

139. Kroes, P. Technical Artefacts: Creations of Mind and Matter: A Philosophy of Engineering Design. Springer (2012). 2.2, II, 14.3, 15, 28, 28.1

140. Kroes, P. Design methodology and the nature of technical artefacts. Design Studies, 23(3), 287-302 (2000).

141. Landin, P. The Mechanical evaluation of Expressions. The Computer Journal, 6 (4): 308-320 (1964). 1.3, 8.2, 10, 11.1

142. Lakatos, I. *Proofs and Refutations*. Cambridge University Press (1976). 14.2

143. Lang, M. Aspects of Mathematical Explanation: Symmetry, Unity and Salience. Philosophical Review, Vol 123, no. 4 (2014). 28.2

144. Leroy, X. Formal Certification of a Compiler Back-End or: Programming a Compiler with a Proof Assistant. ACM SIGPLAN Notices, 41: 42-54 (2006). 24.2

145. Lewis, D. *On the Plurality of Worlds*. Oxford (1986). 21.1

146. Liskov, B., Zilles, S.N. Programming with abstract data types. ACM SIGPLAN Notices, Volume 9 Issue 4. pp 50-59 (1974). 17.2, 21.4

147. Locke, J. *The Clarendon Edition of the Works of John Locke*. Oxford University Press (2015). 2.3, 21.1

148. Luckham, D. Rapide: A Language and Toolset for Causal Event Modeling of Distributed System Architectures, pp. 88-96 (1998). 1.3, 7.1

149. Machennan. J. *Principles of programming languages: design, evaluation and implementation*. Oxford University Press (1999). 19.1, 19.2, 20

150. McConnell, S. *CodeComplete*. Microsoft Press (2004). 17.1, 19.4, 24

151. Machamer, P., Darden; L., Craver, C. F. Philosophy of Science, Vol. 67, No. 1. pp. 1-25 (2000). 28.3

152. McLaughline, P. What Functions Explain – Functional Explanation and Self-Reproducing Systems. Cambridge Studies in Philosophy and Biology. Cambridge University Press (2001). 3.3, 29.1

153. MacQueen, D. Modules for Standard ML, LFP '84. *Proceedings of the 1984 ACM Symposium on LISP and functional programming*: 198-207 (1984). 17.2

154. Magee, J. D. Specifying Distributed Software Architectures. In *Proceedings of 5th European Software Engineering Conference* (1995). 14.1

155. Martin, R. *Agile Software Development: Principles, Patterns, and Practices*. Pearson (2003). 17.5

156. Martin-Löf, P. Constructive mathematics and computer programming. In *Logic, methodology and philosophy of science VI* (Hannover, 1979). Stud. Logic Found. Math., v. 104, pp. 153-175, North-Holland, Amsterdam (1982). 18.6, 26

157. McDowell, J. *Mind, Value, and Reality*. Harvard University Press (2002). 30.5

158. McDowell, J. Are Meaning, Understanding etc. Definite States? in *The Engaged Intellect*, pp. 79-95. Harvard University Press (2009) 30.5

159. McDowell, J. How Not to Read Philosophical Investigations: Brandom's Wittgenstein' in *The Engaged Intellect*, pp. 96-111. Harvard UP (2009) 30.5

160. McDowell, J. Non–Cognitivism and Rule-Following. In *Wittgenstein: to Follow a Rule*. pp. 141-162. S. Holtzman & C. Leich (eds.). Routledge & Kegan Paul (1981) 30.5

161. McDowell, J. Wittgenstein on Following a Rule, Synthese 58, no. 3 (1984). 30.5

162. Meijers, A.W.M. The Relational Ontology of Technical Artifacts. The Empirical Turn in the Philosophy of Technology Research. Philosophy and Technology, Vol 20 (2001). 3.1

163. Michael. M. Occam's Razor. http://michaellant.com/2010/08/10//occams-razor-and-the-art-of-software-design. Last accessed December 31, 2017. 16.2

164. Miller, T. Defining Modules, Modularity and Modularization. Design for Integration in Manufacturing. In *Proceedings of the 13th IPS Research Seminar.* Fuglsoe (1998). 17

165. Miller, A. Rule-Following Skepticism. In Routledge *Companion to Epistemolgy.* S. Bernecker and D. Pritchard (eds.). Routledge (2010). 30.4

166. Milner, R. *Communicating and Mobile Systems: The π-calculus.* Cambridge, UK: Cambridge University Press (1999). 7.1, 8.5

167. Milne, R. and Strachey, C. *A Theory of Programming Language Semantics.* New York, NY: Halsted Press (1977). 2.1, 10

168. Mitchell, J.C. *Foundations for Programming Languages.* MIT Press (1996). 1.3, 2.1, 7.1, 8, 19

169. Morgan, C. *Programming from specifications.* Prentice Hall (1994). 18.5

170. Moor, J.H. Three Myths of Computer Science. The British Journal for the Philosophy of Science, 29(3): 213-222 (1978). 2.2, II, 5

171. Mulmuley, K. *Full Abstraction and Semantic Equivalence.* MIT Press (1987). 10.4

172. O'Keefe, R. *The Craft of Prolog.* MIT Press (1990). 8.3, 8.3

173. Ong, C.H.L. Correspondence between Operational and Denotational Semantics: The Full Abstraction Problem for PCF. In *Handbook of Logic in Computer Science.* pp. 269-356. S. Abramsky, D. Gabbay, T, Maibaum, (eds). Oxford (1995). 10, 10.4, 10.5

174. Oram, A. and Wilson, G. *Beautiful Code.* O'Reilly (2007). 16.1

175. Paquet, J., Serguei, J., Mokhov, A. https://www.academia.edu/357475/. Comparative Studies of Programming Languages. COMP6411 Lecture Notes Revision 1.4. Last accessed December 31, 2017.

176. Parnas, D. L. A Technique for Software Module Specification with Examples. Communications of the ACM, Volume 15. 330-36 (1972). 17.3, 17.7

177. Parnas, D. L. On the Criteria to Be Used in Decomposing Systems into Modules. Communications of the ACM, Volume 15. 1053-58 (1972). 17.7

178. Parnas, D. L., Clements, P., Weiss, D. The Modular Structure of Complex Systems, IEEE Transactions on Software Engineering, SE-11(March): 259-66 (1985) 17.7

179. Parsons, G. *The Philosophy of Design.* Polity (2015). 2.3, 4.3, IV, 15

180. Pattis, R. Pattis Quotes on Programming. http://www.azquotes.com/author/40299-Richard_E_Pattis. Last accessed December 31, 2017. 16.1

181. Pawson, J. *Minimum.* Phaldon (1998). 16.2

182. Pears, D. *Paradox and Platitude in Wittgenstein's Philosophy.* Oxford (2006). 29.2

183. Perry, D. Porter, A. and Lawrence, V. Empirical Studies of Software Engineering: A Roadmap. In *Proceeding ICSE '00 Proceedings of the Conference on The Future of Software Engineering.* pp 345-355. ACM (2000). 13.4

184. Pfleeger, S. and, Atlee, J. *Software Engineering: Theory and Practice.* Pearson (2010). 13

185. Piccinini, G. Computation in Physical Systems. *The Stanford Encyclopedia of Philosophy* (Summer 2015 Edition). 2.4, 4.7

186. Piccinini, G. *Physical Computation: A Mechanistic Account.* Oxford (2015). 28.3

187. Pierce, B.C. *Types and Programming Languages.* MIT Press (2002). 1.3, 19.3

188. Pitowsky, I. Quantum Speed-up of Computations. Philosophy of Science Vol. 69, No. S3 (September 2002), pp. S168-S177. 23.3

189. Plotkin, G.D. A Structural Approach to Operational Semantics (1981). `http://homepages.inf.ed.ac.uk/gdp/` `publications/sos_jlap.pdf`. Last accessed December 31, 2017. 9, 10, 10.3, 19.3

190. Putnam, H. Minds and Machines. In *Dimensions of Mind: A Symposium*. S. Hook (ed.), New York: Collier (1960). pp. 138-164. 2.4, 27.2

191. Quine, Willard van Orman. *Ways of Paradox*. New York: Random House (1966). Cambridge, MA: Harvard University Press (1976). 16.3

192. Rapaport, W. J. Implementation is Semantic Interpretation. The Monist, 109-30 (1999). 11.2, 11.3

193. Rapaport, W.J. Implementation is Semantic Interpretation: more Thoughts. Journal of Experimental & Theoretical Artificial Intelligence (2005). 17(4):385-417. 11.2, 11.3

194. Raymond, E. S. Eric S. Raymond quotes on Programming. `https://en.wikiquote.org/wiki/Eric_S._Raymond`. Last accessed December 31, 2017. 16.1

195. Reynolds, J. C. An Introduction to Polymorphic Lambda Calculus, in *Logical Foundations of Functional Programming*. pp. 77-86. G. Huet (ed). Addison-Wesley (1994). 8.2

196. Richards, M. The BCPL Reference Manual. Memorandum M-352, Project MAC. Cambridge, MA, USA (1967). 8.1

197. Ryle, G. *The Concept of Mind*. The University of Chicago Press (1949). 2.4

198. Sampter, J. Simplicity. `http://www.quotationspage.com/subjects/simplicity/`. Last accessed December 31, 2017. 15.2

199. Schiaffonati, V., Verdicchio, M. Computing and Experiments. Journal Philosophy & Technology, Volume 27 Issue 3. pp 359-376 (2011). 25.4

200. Schildt. H. Java: The Complete Reference. Oracle Publications (2014) 5.2, 8.4

201. Searle, J. (1995). *The Construction of Social Reality*. Penguin (1995). 3.3

202. Shanker, S. G. Wittgenstein versus Turing on the nature of Church's thesis. Notre Dame J. Formal Logic 28, no. 4, 615-649 (1987). 22.5, 22.6

203. Shapiro, E., Sterling, L. *The Art of Prolog: advanced programming techniques*. MIT Press (1994). 8.5

204. D. A. Schmidt. *Denotational Semantics: A Methodology for Language Development*. Allyn & Bacon (1986). (Reprint, 1988). 9, 10, 19.3, 20

205. Schmidt, D. *The structure of typed programming languages*. MIT Press (1994). 20, 20.3

206. Simon, H. A. *The Sciences of the Artificial*. MIT Press (1996). IV

207. Skemp, R. *The Psychology of Learning Mathematics*. Lawrence Erlbaum Associates (1987). Hillsdale, New Jersey, Hove and London. 21.1

208. Sloman, A. Virtual Machines. `https://www.cs.bham.ac.uk/research/projects/cogaff/sloman.virtual.` `slides.ps`. Last accessed December 31, 2017. 11.1

209. Spivey, J.M. *Understanding Z: a specification language and its formal semantics*. Cambridge (1988). 1.3, 12.4

210. Sprevak, M. Three Challenges to Chalmers on Computational Implementation. Journal of Cognitive Science, 13(2), 107-143 (2012). 28.1

211. Steiner, M. Mathematical Explanation. Philosophical Studies, Vol 34, No 2 (1978). 28.2

212. Stoughton, A. *Fully Abstract Models of Programming Languages*. Pitman-Wiley (1988) 10.4

213. Stoy, J. *Denotational Semantics: The Scott-Strachey Approach to Programming Language Semantics*. MIT Press (1977). 1.3, 2.1, 9, 10, 10.4

214. Sommerville, I. *Software Engineering*. 10th Edition. Pearson (2016). 1.3, 6, 13, 13.1, 17.2

215. Shapiro, S. Mathematics and Reality. Philosophy of Science, 50(4): 523-548 (1983). 1, 2

216. Simon, H. *The Sciences of the Artificial*. Third Edition. MIT Press (1996). 3.1

217. Steele, G. *Common Lisp the Language*. Digital Press (1990). 8.2

218. Milne, R. and Strachey, C. *A Theory of Programming Language Semantics*. New York, NY: Halsted Press (1977). 9, 10

219. Strachey, C. Fundamental Concepts in Programming Languages. Higher-Order and Symbolic Computation, 13, pp 11-49. Kluwer Academic Publishers (2000). 9.4, 10, 20

220. Studd, J.P. Abstraction Reconceived. The British Journal of the Philosophy of Science, 67 (2): 579-615 (2016). 21.5

221. Szabó, Z, G. Compositionality. *The Stanford Encyclopedia of Philosophy* (Fall 2013 Edition). 9.3

222. Tedre, M. *The Science of Computing: Shaping a Discipline*. Taylor and Francis (2014). 1, 25.4

223. Tennent, R.D. Language design methods based on semantic principles. Acta Informatica, Volume 8, Issue 2, pp 97-112 (1977). 1.3, 9.3, 20

224. Tennent, R.D. *Semantics of programming languages*. Prentice Hall (1991). 1.3, 2.1, 9, 10

225. Thomasson, A. Artifacts and human concepts. In *Creations of the Mind: Essays on Artifacts and Their Representations*. L. Margolise (ed.). Oxford (2007).

226. Thompson, S. *Haskell: The Craft of Functional Programming*. Addison-Wesley (2011). 5.2, 8.2

227. Thompson, S. *Miranda: The Craft of Functional Programming*. Addison-Wesley (1995). 2.1, 8.2

228. Turing, A. Computing Machinery and Intelligence, Mind, LIX (236): 433-460 (1950). 22.4, 24.3

229. Turing, A. On Computable Numbers, with an Application to the Entscheidungsproblem. *Proceedings of the London Mathematical Society*, Series 2, Vol. 42. pp 230-265 (1936). 8.1, 22.5

230. ACM Turing Award Lectures: The first Twenty Years. ACM Press (2007). 25.4

231. Turner, R. *Constructive Foundations for Functional Languages*. McGrawHill (1991). 18.6

232. Turner, R. Programming Languages as Technical Artifacts. Philosophy of Technology, Volume 27, Issue 3, pp 377-397 (2014). 11.4

233. Turner, R., Angius, N. The Philosophy of Computer Science. *The Stanford Encyclopedia of Philosophy* (Spring 2017 Edition) 1, 2

234. Turner, R. Typed Predicate Logic. http://cswww.essex.ac.uk/staff/turnr/Mypapers/TPLnotes.pdf. Last accessed December 31, 2017. 12, 12.1, 12.5

235. Turner, R. Specification. Minds and Machines, 21(2), 135-152. (2011) 14, 14.1, 14.3, 25.4

236. Turner, R. Understanding Programming Language. Minds and Machines, 17(2): 129-133 (2007).

237. Turner, R. *Computable Models*. Springer (2013). 12, 12.1, 12.2

238. Turner, R. The Ways of Computational Abstraction. https://programme.hypotheses.org 21.7

239. Van Leeuwen, J. *Handbook of theoretical computer science* (vol. A): algorithms and complexity. Springer (1990). 1.2

240. Vermaas, P. E. Ascribing functions to technical artifacts: A challenge to etiological accounts of function. The British Journal of the Philosophy of Science, pp 261-289 (2003). 3

241. Van Vliet, H. *Software Engineering: Principles and Practice* (3rd Edition). John Wiley & Sons (2008). 13

242. Voevodsky, V. Univalent Foundations Project, a modified version of an NSF grant application, pp. 1-12, October, 2010. 25.1

243. Wang, H. *Reflections on Kurt Gödel*. MIT Press (1990). 22.5

244. Warren, D. H. D. An abstract Prolog instruction set (PDF). Menlo Park, CA, USA: Artificial Intelligence Center at SRI International. (1993). 1.3

245. Webopedia. Virtual Machines. http://www.webopedia.com/TERM/V/virtual_machine.html. Last accessed December 31, 2017. 11.1

246. White, G. The Philosophy of Computer Languages. In *The Blackwell Guide to the Philosophy of Computing and Information*. pp. 318-326. L. Floridi (ed.). Blackwell (2004) 10

247. van Wijngaarden, A. Generalized Algol, Symbolic Languages. In *Data Processing*, Proc. Symp. Intl, Computation Center Rome, Gordon & Beach, New York (1962), pp. 409-419. 19.1

248. Virtual Machines. https://www.techopedia.com/definition/4805/virtual-machine-vm. Last accessed December 31, 2017. 11.1

249. Wikipedia. https://en.wikipedia.org/wiki/Encapsulation_(computer_programming) 17.1

250. Wing, J. Computational Thinking. https://www.cs.cmu.edu/~15110-s13/Wing06-ct.pdf. Last accessed December 31, 2017. 1.4

251. Winskel, G. *The Formal Semantics of Programming Languages: An Introduction*. MIT Press (1993). 9, 10

252. Wirth, N. On the Design of Programming Languages. *Proc. IFIP Congress 74*, 386-393. (1974). 16

253. Wirth, N. The Programming Language Pascal. Acta Informatica, Volume 1, pp 35-63 (1971). 8.1, 18.3

254. Wirth, N. MODULA: language for modular multiprogramming. *Eidgenössische Technische Hochschule Zürich* (1976). 8.1

255. Wittgenstein, L. *Philosophical Investigations*. Blackwell (1953). 22.7, 29.1, 30.2, 30.5

256. Wittgenstein, L. *Wittgenstein's Lectures on the Foundations of Mathematics*. Cora Diamond (ed). The Harvester Press, Sussex (1976). 14.2, 22.7, 25.1

257. Wittgenstein, L. *Remarks on the Foundations of Mathematics*. G. H. von Wright, R. Rhees, and G. E. M. Anscombe (eds.). Oxford (1978). 29, 30.2

258. Woodcock, J. A. *Using Z- Specifications, Refinement and Proof*. Prentice Hall (1996). 1.3, 2.3, 7.1, 12, 12.4

259. Wright, C. Is Hume's Principle Analytic? Notre Dame J. Formal Logic, 40(1), pp.6-30 (1999). 21.2

260. Wright, C. *Frege's Conception of Numbers as Objects*. Aberdeen University Press (1983). 2.3, 21.2

Index

Printed in the United States
By Bookmasters